THE COMPLETE FIRE-FIGHTER'S EXAM PREPARATION BOOK

EVERYTHING YOU NEED TO KNOW THOROUGHLY COVERED IN ONE BOOK

Complete Practice Exams -- Answer Keys -- Self-scoring Tables
Proven Tips for Boosting Scores -- Passing the Physical Exam
Memory Aids to Help You Master the Recall Test
PLUS: Everything You Need to Know About Firefighting as a Career

NORMAN HALL
ILLUSTRATOR: SHANNON HALL

Copyright ©1992, Norman S. Hall. All rights reserved.
No part of this work may be reproduced in any form, or by any means,
without the permission of the publisher. Exceptions are made for
brief excerpts to be used in published reviews.

Published by Adams Media Corporation
260 Center Street, Holbrook, MA 02343

ISBN: 1-55850-052-9

Printed in Canada

H   I   J

This publication is designed to provide accurate and authoritative information with regard to the subject matter covered. It is sold with the understanding that the publisher is not engaged in rendering legal, accounting, or other professional advice. If legal advice or other expert assistance is required, the services of a qualified professional person should be sought.
— From a *Declaration of Principles* jointly adopted by a Committee of the American Bar Association and a Committee of Publishers and Associations.

This book is available at quantity discounts for bulk purchases.
For information, call 1-800-872-5627 (in Massachusetts, 781-767-8100).

Visit our home page at http://www.adamsmedia.com

3 2872 50108 3848

TABLE OF CONTENTS

INTRODUCTION 5
   General Job Description, Test-Taking Strategies, Study Suggestions, Exam Content 8

MEMORY RECALL 9
   Names, Numbers, Floor Plans
      Sample Floor Plans 13
   Descriptive Passages
      Sample Descriptive Passages 25
      Answers to Sample Questions 33

MECHANICAL APTITUDE 34
   Tools
      Sample Questions 50
      Answers to Sample Questions 58
   Mechanical Principles, Simple Machines
      Sample Questions 69
      Answers to Sample Questions 79

DIRECTIONAL ORIENTATION 85
   Sample Questions 87
   Answers to Sample Questions 105

MATHEMATICS 109
   Mathematical Principles
      Sample Questions 117
      Answers to Sample Questions 123

READING COMPREHENSION 130
   Sample Questions 136
   Answers to Sample Questions 145

JUDGMENT AND REASONING 148
   Sample Questions 148
   Answers to Sample Questions 155

PRACTICE EXAM 1 159
   Exam 1 Answer Sheet (181) and Answers 183

PRACTICE EXAM 2 195
   Exam 2 Answer Sheet (213) and Answers 215

PRACTICE EXAM 3 227
   Exam 1 Answer Sheet (251) and Answers 253

PRACTICE EXAM 4 265
   Exam 1 Answer Sheet (287) and Answers 289

PHYSICAL FITNESS 301

THE INTERVIEW—THE FINAL PHASE OF SELECTION 307

PREFACE

Congratulations on taking the most important step toward achieving a career in firefighting. Your purchase of this study guide demonstrates your determination to use the best preparation material available. Considering the intense competition for career firefighting jobs, it is imperative that you achieve a high score on the written exam. Thorough preparation is the key.

This book examines the various sections of the written exam and provides helpful insights and tips to guide you through the more difficult parts. Extensive research went into making this study guide. Firefighter exams given in various regions around the country were scrutinized and then compared and analyzed. One inescapable fact that became evident early on is that no two exams are exactly alike: Some city and county fire departments have their own test files, while others contract testing out to private firms that specialize in screening applicants. Despite the wide array of material encountered on various exams, this book offers an excellent overview of what will be expected of you on an actual exam. In fact, most questions in this book closely parallel those seen on past exams.

The second part of the screening process is the physical fitness test. Normally this is given to applicants who rank in the top 20 to 30 percent on the written exam. Similar to the written exam, the physical fitness exam has been conducted in a number of different forms. The final section of the book has been dedicated to this subject, and includes some suggestions that will help you pass. (Physical fitness exams are graded on a pass/fail basis only).

After studying the material in this guide, you will be able to approach the written portion of the exam with a sense of confidence and ease. In addition, this study guide offers an assurance unmatched by any other publication: Guaranteed test results. If you do not score 80 percent or better on the written exam, the purchase price of this study guide will be refunded in full by the publisher. (See the last page of this book for further details.)

Before you begin your studies, I would like to wish you the best in your future job endeavors. Once you are hired in the fire department, the rewards and job satisfactions are great and the service you provide to the local community invaluable.

— *Norman S. Hall*

INTRODUCTION

There are an estimated 300,000 career firefighter positions in the United States. Most of these are available within fire departments in the larger municipalities, which have a substantial tax base for funding fire-protection services. Smaller towns and rural communities must usually rely solely on volunteer firefighting services.

An annual personnel turnover rate of 10 to 15 percent is typical. This rate can be attributed to retirements, promotions, disabilities, transfers, and unspecified personal reasons. Often fire departments operating in areas of population growth are appropriated extra funds for expansion. Conversely, some cities may experience a contraction of population because of regional economics. In such cases, new positions will not be created. However, layoffs in this profession are relatively rare.

Because of normal attrition and potential community growth, it is important that fire departments maintain active registers of qualified applicants to fill vacancies as they arise. The frequency of exams can vary. Ordinarily, exams are given yearly. Public announcements are made several weeks in advance of the exam in the local media (usually newspapers and radio) specifying when and where individuals can apply for the test.

However, it is strongly recommended that you fill out a job application form now. Application forms can be obtained from your local fire department or civil service bureau. Then, when exam dates are determined, you will be notified by mail. By applying now, you will eliminate the possibility of missing an exam or exam announcement and have to wait another year for the next exam.

There are some basic minimum requirements outside the written exam that job applicants must meet for employment eligibility. You must

- possess a high school diploma or G.E.D.
- be at least eighteen years of age
- have a valid state drivers license and a reasonable driving record
- be free of any felony convictions and not have received a dishonorable discharge from the military
- prove your residency (in some instances)
- be able to pass a medical and physical exam

Other desirable abilities that would better qualify an applicant for the job would include

- good communication skills (oral and written)
- strong mechanical aptitude
- ability to work under pressure and maintain a collective sense of direction
- some knowledge of first aid and emergency medical care
- ability to work with others as a unit and follow instructions of superiors
- self-motivation and a willingness to work hard

GENERAL JOB DESCRIPTION

The working conditions encountered by firefighters in the line of duty merit consideration, too. Since two situations are never exactly alike, this demands flexibility. Duties can be as mundane as regular cleanup and maintenance of station equipment or can involve such physical hazards as working at great heights, exposure to fire and temperature extremes, smoke inhalation, and falling debris, and dealing with unstable or toxic chemicals and gases. As an emergency medical technician (EMT) or paramedic, you can be exposed to victims who have suffered severe injuries (burn trauma, stroke, heart attack, wounds with profuse bleeding or internal hemorrhaging) or have communicable diseases. Life and death decisions can hang in the balance purely on the training and expertise of the personnel who respond to an alarm. These conditions can arise at any time of the day or night and can be made worse by prevailing weather conditions. Nevertheless, a firefighter must be ready and willing to respond to any situation immediately.

Work schedules vary from area to area, but firefighters normally spend twenty-four hours on duty and forty-eight hours off. Some departments have working hours comparable to other jobs (for example, eight- to ten-hour daily shifts). Whatever the format, forty-five to fifty work hours per week is standard, with the expectation of overtime if an emergency so demands. For stations that use the twenty-four-hours-on, forty-eight-hours-off schedule, firefighters are expected to live in quarters provided at the station. Amenities usually include a kitchen, showers, locker room, study, and sleeping areas. The entire shift, however, is not completely dedicated to working or sleeping. Time is provided for leisure or technical reading, games, and other activities, with the provision that they be done within the perimeter of the station for reasons of readiness.

General job duties for entry-level firefighters begin with extensive training and study. New recruits typically engage in one to three months of intense study of firefighting technology and are made aware of the applications and uses of all equipment. During a probationary period, new hires are given the opportunity to demonstrate their learned skills while reacting to real situations, usually under the close supervision of a fire lieutenant.

Other responsibilities include regular equipment maintenance and station cleanup, participation in training drills and established physical fitness programs, assisting in fire-safety inspections, determining compliance with local codes and filing reports and citations accordingly, performing salvage operations such as sweeping water and removing debris, transporting injured or sick patients to local emergency facilities, filing detailed reports of such incidents to assist medical personnel in evaluating patients, and speaking in schools or community assemblies about fire safety and evacuation procedures.

Promotions are normally based on three different aspects of a firefighter's performance. First and most important is the quality of work performed by the applicant and the degree of enthusiasm with which it is conducted. Second is performance on written examinations, held periodically to determine the level of training or competence an applicant possesses. The higher the test score, the better the chances are for promotion. The last qualification for promotion is seniority and high job performance; those who have demonstrated high capabilities in their job performance and have done so for a significant time may be considered for promotion. Each fire department weighs these considerations differently. Ultimately, the most qualified people with the dedication to improve themselves are the ones most likely to advance.

Earnings are directly related to the level of training an employee receives. Entry-level firefighters earn on average approximately $2,200 per month, excluding overtime. Fire lieutenants earn an average of $2,600 per month. Those at the rank of battalion chief or higher can demand a salary of between $40,000 to $50,000 per year. Incentive bonuses in the amount of an extra $100 to $400 per month are sometimes given to firefighters with EMT or paramedic certification.

Most fire departments offer a host of excellent benefits, including medical and, in some cases, dental insurance; paid vacation and sick leave; and a solid retirement plan and pension. Clothing allowances are also provided to cover the expense of buying uniforms and related clothing. Further information about wages and benefits can be acquired by contacting the fire department to which you are applying.

TEST-TAKING STRATEGIES

You may have a firm grip of the subject matter on the test, but if you make mistakes filling in the answer sheet, the resulting test score will not truly reflect your knowledge. It is not uncommon for applicants to discover that they received less than a passing score of 70 percent or better because a few simple rules weren't followed.

When the examiner gives any kind of instruction, pay attention. The examiner will explain how properly to fill in personal information that will be used to identify your exam results. He or she will further explain how and when to proceed on the exam. Do not deviate in any manner from the established test procedure. If you do, you may disqualify yourself altogether.

INTRODUCTION

All of the exams scrutinized to compile this study guide were of the multiple-choice variety. This immediately makes your job a bit less difficult; you know that one of the choices has to be correct. Even if the answer isn't immediately apparent, some choices can be eliminated, further increasing your chances of selecting the right answer. Another advantage of multiple choice exams is that if the time remaining to complete the exam becomes a factor, you can always mark answers at random and still have a 25 percent chance of picking the correct choice. You will not be penalized for wrong answers. Consequently, it is better not to leave any questions unanswered, if possible.

Most of the exams seen include one hundred questions and take approximately two hours to complete. The answer sheet is usually separate from the test booklet, and a sheet of scratch paper for mathematical calculations and general figuring is provided. It is important not to make any extra marks on the answer sheet, because the machine that scores the exams can misconstrue a mark as an incorrect answer. An example of an answer blank is provided below to demonstrate how to mark an answer properly.

1. (A) ● (C) (D)   3. ● (B) (C) (D)   5. ● (B) (C) (D)
2. (A) (B) (C) ●   4. (A) (B) ● (D)   6. (A) (B) ● (D)

The following examples are answer blanks that have been improperly marked, leading to a poor test score.

1. (A) (B) (✓) (D)   3. (A) (B) (C) (D)   5. (A) (B) (∅) (D)
2. (A) (B) (✗) (D)   4. (A) (B) (C) (D)   6. (A) (B) (C) (Ⓓ)

If you change your mind about any answer, be certain to erase the original answer completely. If two answers are apparently marked the scoring machine will consider the answer incorrect.

One costly mistake is marking answers that do not correspond to the question you are working on. As a simple rule, check every ten questions or so to make sure you are marking the matching answers for the corresponding question.

When you look over test questions, be sure to read them carefully and completely. It is easy to fall prey to reading the first or second choice and selecting one of them as the correct answer without bothering to examine the remaining options. Read the entire passage thoroughly and then mark your answer sheet accordingly. Pay close attention to conjunctions such as *and, but, or, when, if, because, though, whereas*, and *besides*. These key words can completely alter the meaning of a question. If you overlook one of these words, there is a good chance that you will select the wrong answer. It can't be emphasized enough to read the entire passage closely before even thinking about the answer sheet.

Do not spend too much time on a question about which you are unsure. Make an educated guess, or skip that question and return to it when you have finished the others. As a timesaver, you can cross out in the test booklet options that you know are wrong. When you return to that question, you can quickly focus on the remaining selections. If you skip a question, be sure to skip the corresponding answer blank, too.

As a rule, the first choice you select is the correct one. Statistics show that answers changed are typically wrong. All too frequently, people read too much into a question and obscure the proper choice. Mark the first answer that seems apparent.

It is important to get plenty of rest the night before the exam. Staying up late and reviewing all that you have studied is actually counterproductive. "Cramming" does not improve test performance and can increase your own nervous tension.

Give yourself plenty of time to arrive at the test site; twenty to thirty minutes early is desirable. Unforeseen events could detain you at home, or traffic congestion en route to the exam could make you late. Once the exam has been started, latecomers will not be permitted to take the test.

Before you arrive to take the test, be sure to have some form of identification such as a valid driver's license, a validated test application form, and a couple of No. 2 pencils. Calculators, sliderules, and notes will not be permitted in the examination room.

STUDY SUGGESTIONS

This is not the kind of exam on which you can hope for a high test score after just cramming for the night before. Regular study times should be established and tailored to your comfort. Each person's schedule is different. Some people prefer to study for one or two hours at a time with intermittent breaks, while others prefer several hours of straight study. Regardless of how you study, it is important that you do it regularly, and not rely on a marathon. You will remember the subject matter more easily and comprehend it better by establishing regular study habits.

Where you study is important, too. Eliminate any distractions that can disrupt your studies. The television, the telephone, and children can hinder quality study time. I suggest you set aside one of the rooms in your home as a study place and use it to isolate yourself from distractions. If you elect to use a bedroom as a study area, avoid lying in bed while you read. Otherwise, you may find yourself more inclined to sleep than to learn. It is important to have a good desk, a comfortable chair, and adequate lighting; anything less can hamper studying. If studying in your home is not feasible, go to your local library or some other place that offers an environment conducive to study.

Again, be sure to get plenty of rest. It is counterproductive and slows learning if you try to study when you are overly tired. It is also important not to skip meals. Your level of concentration during the exam can suffer if you lack proper nutrition. Coffee and or other stimulants are not recommended.

Good study habits can have a significant impact on how well you do on the exam. If you follow these few simple guidelines, you can approach the exam more relaxed and confident, two essential ingredients for top performance on any exam.

EXAM CONTENT

The firefighter exam is a general aptitude test. However, most of the questions on it pertain to firefighting or some related field. You are not expected to have the same level of knowledge as an experienced firefighter. Instead, your grasp of general concepts, reasoning, and logic are the main focus of the exam. If you do have the time to read some supplementary firefighting literature, you are encouraged to do so. Any further broadening of your knowledge will be beneficial.

The firefighter test itself can be broken down into six topics of study:

1. Memory recall
2. Mechanical aptitude
3. Directional orientation
4. Basic mathematics
5. Reading comprehension
6. Judgment and reasoning

Each of these areas of study will be discussed at the beginning of this study guide and then followed by sample test questions and answers. Test strategy and hints will also be elaborated on as they apply to each subject area.

It is important to note that there can be significant variation in test content, judging by what was seen on past exams from around the country. Some tests place a stronger emphasis on one or more subject areas while having few questions, if any, relating to other topics. To prepare yourself adequately for such a test, all six areas of study need equal consideration. Then you will be prepared regardless of what you may encounter on your exam.

Following these studies are four practice examinations. Approach each as though you were taking the real exam. Apply the test-taking strategies previously discussed and give yourself two hours of uninterrupted time for each exam. The more closely you can simulate actual test conditions, the better. By getting a realistic "feel" for the actual exam, you can reduce test anxiety.

MEMORY RECALL

Having a good memory has many advantages, but in the line of duty for a firefighter it can make a pivotal difference in how an emergency situation is handled. A firefighter who is quick to grasp floor plans detailing exits, fire hose connections, and the locations of fire-suppression systems, can handle emergencies more effectively and safely. This can be particularly important when structures being entered are partially engulfed in smoke. An incomplete understanding of a building's layout may lead to confusion, hampering rescue operations or effective fire suppression.

Another example involves directives issued by a commanding officer at the scene of an emergency. It is imperative that orders be followed quickly, accurately, and completely, without having to be repeated. Valuable time could be lost, and the effectiveness of an operation could suffer, if commands are second-guessed or forgotten. If a firefighter has a good memory, his or her actions will reflect a higher degree of competence and a more positive attitude toward any challenge.

The memory recall section was placed at the beginning of this study guide because it is normally the first subject encountered on the actual exam. Test examiners prefer to arrange an exam in this manner so that the rest of the exam can be given without any further interruptions. Typically, a second exam booklet, a key, or some other form of diagram is passed out to test applicants. Applicants are allowed an allotted period of time to memorize as much of the diagram as possible; then the material is collected. The question and answer sheet for this section may be handled separately, but are more than likely to be an integral part of the main exam. In any case, the key or diagram is not available for reference during the test. All your answers must be arrived at by memory alone.

Memory recall tests usually contain information on floor plans, sketches of emergency scenes, or descriptive passages. Ordinarily, you are given five minutes to study what is presented. The memory test may appear simple at first glance, but may require memorizing up to fifty different items.

Unless you are gifted with a photographic memory, memorizing that amount of material in a short time is almost impossible. Don't despair. There is a technique that can be of invaluable assistance in aiding your memorization, a technique called *imagery and association*. Any memory task can be simplified by using this system. It requires you to form images in your mind relevant to the item to be memorized. Then you link these images by association.

A. NAMES

Street names are some of the items that must be committed to memory for the exam. Use the following street addresses as examples:

   Jorganson Street
   Phillips Avenue
   Tremont
   Tricia
   Edgewater Boulevard
   Bloomington

Most people would approach this exercise by rote memorization, or in other words, repetition of thought until recall can be accomplished. This is a boring way of doing things, wouldn't you say? Believe it or not, you can actually have a little fun at doing memory exercises. Now, using imagery and association, look at those same street names again and see what key word derivatives have been used and what images we can associate with them.

COMPLETE FIRE-FIGHTERS EXAM PREPARATION BOOK

For example:

Jorganson Street — Jogger
Phillips Avenue — Phillips screwdriver
Tremont — Tree
Tricia — Tricyle
Edgewater — Edge
Bloomington — Blossoms

Carry the process one step further and place those key word derivatives in a bizarre context or situation. We have developed the story below:

A *JOGGER* with his pockets completely stuffed with *PHILLIPS SCREWDRIVERS* wasn't paying attention and ran into a giant *TREE*. After dusting himself off, he jumped on a child's *TRICYLE* and pedaled it to the *EDGE* of a pool filled with flower *BLOSSOMS*.

Sounds ridiculous, doesn't it? However, because of its strong images, you will not easily forget this kind of story.

Another advantage of this technique is that you can remember items in their respective order by simply reviewing where they fit in the story in relation to the other items.

Look at the list of street names below and develop a story using this technique. There are no right or wrong key word derivatives. What is important is that the images conjure up a clear picture in your mind and then interlink.

Work on each of these columns separately.

Bedford Ave.	Apple Dr.	Anderson Blvd.	Bayberry Rd.
Wellington	Constantine Way	Cannon Ave.	Hickory Ridge Dr.
Walker St.	Bristol	Foxtail Run	Ebony Ln.
Penny Ln.	Echo Ave.	Arsenal Way	Ester Ct.
Ridgemont Dr.	Darrington	Jacobson St.	Steinbald Ln.
Bowmont	Smalley St.	Prince Williams	Georgia St.

Once you have finished this exercise, cover the addresses just memorized and see if you can remember all twenty-four items. If your four stories are bizarre enough, you certainly can have this entire list committed to memory in a short time.

B. NUMBERS

Numbers are another problem in memory recall. For most people numbers are difficult to memorize because they are intangible. To simplify this problem, numbers can be transposed into letters of the alphabet so that words can be formed and associated accordingly. Below is the format for transposition. Remember this format as if it were your own Social Security number because in the future you will draw from it regularly.

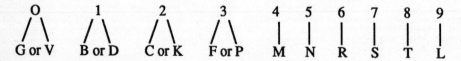

(All other letters can be incorporated into words without bearing any significance.)

For instance, let's say you are given the number 10603328157. Memorizing this number so well that you can recall it after any length of time would be very difficult. However, by using this memory system, you could use the number to spell out a variety of memorable things. (Here is your chance to use your creativity!) After you have had the chance to figure out what words can code such a number, one particular problem should become apparent; the more numbers you try to cram into one word, the harder it is to find a compatible word in the English vocabulary. To simplify matters, there are two alternatives from which to choose. The first is to take two numbers at a time, form a word and associate it with the next word. Dealing with the same number (10603328157) DOG could be derived from the number 10, RUG from 60, PIPE from 33, CAT from 28, BONE from 15 and S from 7. There are many ways you could imagine and link these words. One possibility would be: a DOG lying on a RUG and smoking a PIPE while a CAT prances by carrying a BONE shaped like an S. That is just one solution to memorizing that long number. Other words and stories could work just as well.

The second alternative, which offers greater flexibility, is using words of any length but making only the first two significant letters of the word applicable to your story. For example, the word DIG/GING could represent 10 in the number *10*603328157.

RAV/EN = 60 or RUG/BY = 60 or REV/OLVER = 60
POP/ULATION = 33 or PUP/PY = 33 or PEP/PER = 33
CAT/ERPILLAR = 28 or CAT/TLE = 28 or COT/TON = 28
BIN/OCULAR = 15 or BEAN/S = 15 or DIN/NER = 15

By doing this, you have a larger number of words at your disposal to put into stories. With a little originality, it can be fun to see what you can imagine for any number given.

Below are exercises to help you apply this system. The first group of numbers are meant as a transposition exercise. See how many different words you can use to represent each number. The second series is for practice with transposition and story fabrication. This technique may seem difficult at first, but with practice it will enhance your memory capabilities tenfold.

I.
44	63	86	40
53	97	93	32
61	10	48	26
13	3	60	91
12	57	35	99
8	52	27	16
41	11	21	68

II.
1754732115810   63211347890
6980421569497   145344175328
147329944710   917403218977
8321355572119   638146119900
488770509453   433351896487
1530197865321   765320146991

C. FLOOR PLANS

When you are presented with a floor plan or sketch of an emergency situation on your exam, try to mentally walk your way through the diagram. Pay particular attention to details. For example: Is there an emergency caused by fire or smoke, and if so, where does it originate and what can it imperil? Are there any people or pets present? Where are the exits, windows, fire alarms, etc., and are there any apparent obstructions that could block a potential escape? Street names and house numbers merit attention, too. Let's look at an example and a story that will help us remember necessary details.

EXAMPLE

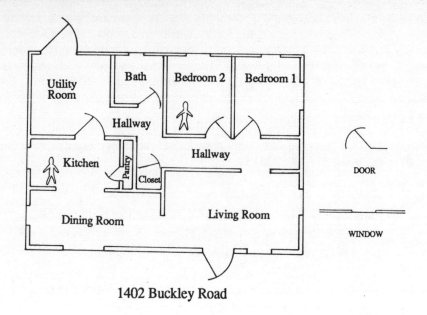

1402 Buckley Road

Focusing on the address first, *1402 BUCKLEY RD* can be transposed into *DEMO*nstration (14) model of a *VAC*uum (02) cleaner *BUCK*ing (*BUCKLEY*) wildly. Despite the antics of this appliance, we enter the front door to the home, and the first thing we notice is the *LIVING ROOM*. *TELEVISIONS* are commonplace in living room areas. What is implicit with a TV set? An *ANTENNA*, of course. However we will think of rabbit ears instead. So here sits an enormous rabbit in the middle of the living room. As a symbol of the number of *WINDOWS* in the living room, think of *TWO* spotlights converging on the rabbit. To the left, we enter the *DINING ROOM*, complete with table and place settings. However, only piles of knives and spoons are apparent. This again is a symbolic attempt to indicate that there are only *TWO WINDOWS* present. We next go to the *KITCHEN* where we notice that an individual has wedged a bar between the window and the pantry door. This was a vain attempt to prevent all of the food from bursting out of the *PANTRY*. Next, we enter the *UTILITY ROOM*. This is properly named because the entire floor is covered with utility knives. The *BACK DOOR* can be opened, but only after kicking the knives aside. As we proceed down the hallway, a small room, presumably the *BATHROOM*, is filled to the ceiling with toilet paper. A single roll of toilet paper (again symbolic of a *SINGLE WINDOW*) rolls out of the bathroom and abruptly disappears in the *CLOSET* across the hall. Continuing further down the hallway and to our left is *BEDROOM 2*. Upon opening the door we see someone squeezing both of his feet into a single shoe (again, symbolic of a single window). We proceed to *BEDROOM 1* and notice *TWO WINDOWS* shaking hands with one another. We quickly leave the premises to preserve our sanity. We look into the living room and there still sits that enormous rabbit with two spotlights on it.

As ridiculous as this story sounds, it covers virtually everything that needs to be memorized. The layout of the home is understood, the location of its residents is established, exits and windows are accounted for, and the address of the premises is identified. Again, any number of stories would have worked. What is important is that all pertinent details be incorporated into your story.

Be careful not to get carried away. Keep your story as short and simple as possible. There is enough information to memorize without complicating the task with unnecessary items.

MEMORY RECALL

SAMPLE FLOOR PLAN I
Study the floor plan below for five minutes. *DO NOT* exceed the time allowed; if you do, you will forfeit the true sense of how an exam actually works. When time is up, cover the key and proceed to the questions given.

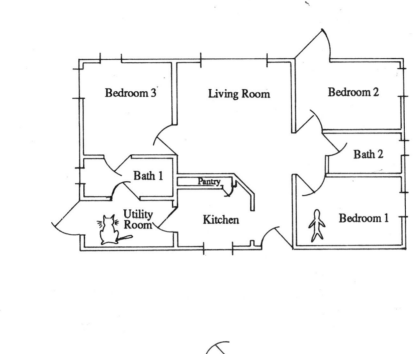

WINDOW   DOOR   DOORWAY

2511 Orchard Lane

1. How many doors lead to the immediate exterior of this structure?
    A. 1　　　　B. 2　　　　C. 3　　　　D. 4

2. Which bedroom(s) has/have a master bathroom?
    A. 1　　　　B. 2　　　　C. 3　　　　D. 1 & 3

3. How many bedroom windows are there?
    A. 3　　　　B. 4　　　　C. 5　　　　D. 6

4. What is the street address of this residence?
    A. 2155 Orchard Lane
    B. 2151 Orange Lane
    C. 2011 Orange Lane
    D. 2511 Orchard Lane

5. If a person were resting in Bedroom 3 and a fire started in the living room, what would be the safest means to evacuate the house?
    A. Through the master bathroom and out the door in the utility room.
    B. Through the living room to the main entry foyer.
    C. Through the living room and into Bedroom #1.
    D. Through the kitchen to the main entry foyer.

6. Which of the bedrooms has/have an exit to the exterior of the house?
    A. 1 & 2　　　　B. 3　　　　C. 3 & 1　　　　D. 2

7. Where would be the worst possible place for someone to seek refuge from fire and smoke?
    A. Living room
    B. Bathroom
    C. Bedroom 1 & 3
    D. Pantry

8. According to the floor plan, which room has no windows at all?
    A. Utility room
    B. Bathroom 1
    C. Bedroom 1
    D. Bathroom 2

9. How many people were shown in the home?
    A. 4　　　　B. 3　　　　C. 2　　　　D. 1

10. Which room had 2 windows and 2 doors?
    A. Bedroom 1
    B. Bedroom 3
    C. Living room
    D. None of the rooms shown

11. Where is the family pet located in the drawing?
    A. Bedroom 3
    B. Pantry
    C. Utility room
    D. Living room

MEMORY RECALL

12. Which of the following statements is true?
    A. Bedroom 3 has more exits than bedroom 2.
    B. The kitchen can be accessed via the main entry foyer and the utility room.
    C. The master bathroom connects to both bedroom 1 and bedroom 2.
    D. The living room is the smallest room in the house.

13. If there were a fire in Bedroom 1, how many potential avenues are there to attack the fire with a hose of pressurized water?
    A. 4 B. 3 C. 2 D. 1

14. If fire alarms were installed in all of the bedrooms and the living room, which one would probably respond first to smoke originating from a kitchen appliance?
    A. Alarm in the utility room
    B. Alarm in Bedroom 2
    C. Alarm in Bedroom 1
    D. Alarm in the living room

15. If Bedroom 2 was considered the northeast corner of the house, where is the utility room in relation to Bedroom 2?
    A. Southwest B. North northeast C. Southeast D. Northwest

COMPLETE FIRE-FIGHTERS EXAM PREPARATION BOOK

ANSWER SHEET FOR SAMPLE FLOOR PLAN I

1. Ⓐ Ⓑ Ⓒ Ⓓ
2. Ⓐ Ⓑ Ⓒ Ⓓ
3. Ⓐ Ⓑ Ⓒ Ⓓ
4. Ⓐ Ⓑ Ⓒ Ⓓ
5. Ⓐ Ⓑ Ⓒ Ⓓ

6. Ⓐ Ⓑ Ⓒ Ⓓ
7. Ⓐ Ⓑ Ⓒ Ⓓ
8. Ⓐ Ⓑ Ⓒ Ⓓ
9. Ⓐ Ⓑ Ⓒ Ⓓ
10. Ⓐ Ⓑ Ⓒ Ⓓ

11. Ⓐ Ⓑ Ⓒ Ⓓ
12. Ⓐ Ⓑ Ⓒ Ⓓ
13. Ⓐ Ⓑ Ⓒ Ⓓ
14. Ⓐ Ⓑ Ⓒ Ⓓ
15. Ⓐ Ⓑ Ⓒ Ⓓ

ANSWERS CAN BE FOUND ON PAGE 33.

SAMPLE FLOOR PLAN II
Study the diagram shown below for five minutes. When your time is up cover the key and proceed to the questions given. DO NOT TURN BACK TO THIS PAGE FOR FURTHER REFERENCE.

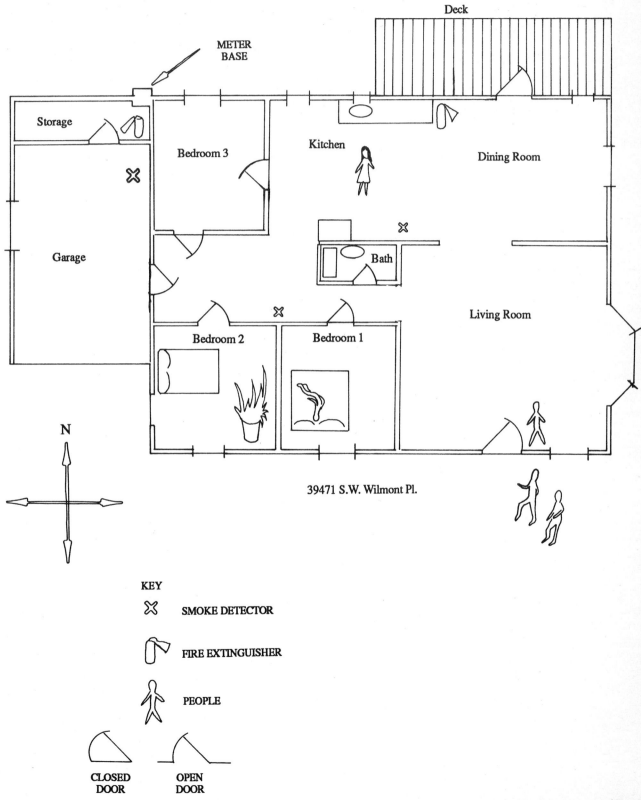

1. How many smoke detectors and fire extinguishers are apparent in this diagram?
    A. 3 and 1, respectively
    B. 4 and 2, respectively
    C. 3 and 2, respectively
    D. 2 and 3, respectively

2. What is the street address of this residence?
    A. 34971 SE Wilmont Avenue
    B. 39417 SW Wilmont Place
    C. 99417 NW Wilmont Avenue
    D. 39471 SW Wilmont Place

3. This structure is more than likely a
    A. Warehouse for storage
    B. Single-family detached ranch home
    C. Multilevel apartment building
    D. Aggregate condominium

4. The fire is located at what side of the house?
    A. Southwest   B. Southeast   C. Northwest   D. Northeast

5. How many people are inside the house, according to the diagram?
    A. 4   B. 3   C. 2   D. 1

6. If electrical power had to be shut off, where would the fuse box or circuit breaker most likely be located?
    A. Bedroom 2
    B. The kitchen
    C. The exterior wall of the dining room
    D. The storage room

7. Which room or rooms has/have only one door closed?
    A. Bedroom 1
    B. Bedroom 2
    C. Bedroom 1 and the bathroom
    D. All doors throughout the home were closed

8. Obviously, the fire poses a threat to everyone in the household, but which person is in the most immediate danger?
    A. The person resting in Bedroom 1
    B. The person in the living room
    C. The person resting in Bedroom 3
    D. The person in the kitchen

9. Which room in the house has a bow-bay window?
    A. Dining room
    B. Bathroom
    C. Kitchen
    D. Living room

MEMORY RECALL

10. The outdoor deck is on what side of the house?
    A. West B. Northeast C. Southeast D. Southwest

11. How many doors and windows does this structure have?
    A. 9 windows and 14 doors
    B. 11 windows and 10 doors
    C. 13 windows and 9 doors
    D. 10 windows and 13 doors

12. Where is the fire in relation to Bedroom 3?
    A. South B. East C. North D. West

13. Where is the garage in relation to the bathroom?
    A. Southwest B. North C. West D. Northeast

14. Assuming the fire were extinguished quickly, sparing the house any real structural damage, what room would suffer the least smoke damage?
    A. Kitchen
    B. Bathroom
    C. Dining room
    D. Storage room

15. How many people were in the dining room at the time of the fire?
    A. According to the floor plan, the dwelling has no dining room
    B. 1
    C. 2
    D. 3 people were in the house, but no one was in the dining room during the fire

COMPLETE FIRE-FIGHTERS EXAM PREPARATION BOOK

ANSWER SHEET FOR SAMPLE FLOOR PLAN II

1. Ⓐ Ⓑ Ⓒ Ⓓ   6. Ⓐ Ⓑ Ⓒ Ⓓ   11. Ⓐ Ⓑ Ⓒ Ⓓ
2. Ⓐ Ⓑ Ⓒ Ⓓ   7. Ⓐ Ⓑ Ⓒ Ⓓ   12. Ⓐ Ⓑ Ⓒ Ⓓ
3. Ⓐ Ⓑ Ⓒ Ⓓ   8. Ⓐ Ⓑ Ⓒ Ⓓ   13. Ⓐ Ⓑ Ⓒ Ⓓ
4. Ⓐ Ⓑ Ⓒ Ⓓ   9. Ⓐ Ⓑ Ⓒ Ⓓ   14. Ⓐ Ⓑ Ⓒ Ⓓ
5. Ⓐ Ⓑ Ⓒ Ⓓ   10. Ⓐ Ⓑ Ⓒ Ⓓ   15. Ⓐ Ⓑ Ⓒ Ⓓ

ANSWERS CAN BE FOUND ON PAGE 33.

MEMORY RECALL

SAMPLE SKETCH I

Study the sketch below for five minutes. When your time is up, cover the sketch and answer the questions given. DO NOT TURN BACK TO THIS PAGE FOR FURTHER REFERENCE.

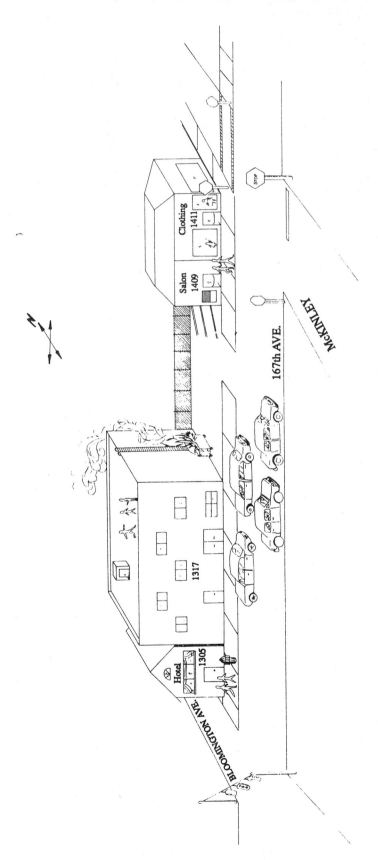

1. What is the address of the tenant most likely to own the dumpster that is the focus of the emergency?
    A. 1303 167th Avenue
    B. 1307 167th Avenue
    C. 1307 Bloomington Avenue
    D. 1317 167th Avenue

2. Judging from what was seen in the sketch, it can be assumed that:
    A. 167th Avenue is a one-way street.
    B. McKinnley is a one-way street.
    C. 167th Avenue is a two-way street.
    D. 167th Avenue is a two-way street that intersects Bloomington Avenue and McKinnley, both of which are one-way streets.

3. At the time of the fire, what was the direction of the wind?
    A. Northwesterly
    B. Southwesterly
    C. Southerly
    D. Wind direction cannot be determined

4. Where is the fire hydrant located in the sketch?
    A. In front of the building at 1305 167th
    B. On the northwest corner of Bloomington Avenue and 167th
    C. In the parking lot east of the warehouse
    D. On the corner of McKinnley and 167th

5. How many people are on the roof of the warehouse?
    A. 2   B. 3   C. 4   D. 5

6. In what direction are parked vehicles facing on the south side of 167th Avenue?
    A. Southwest   B. North   C. East   D. West

7. Which building has an emergency fire escape on its west wall?
    A. 1307 167th Avenue
    B. 1317 167th Avenue
    C. 1401 167th Avenue
    D. No building in the sketch has a visible west-wall fire escape.

8. How many pedestrians are in front of the hotel?
    A. 1   B. 2   C. 3   D. 4

9. How many vehicles are parked on 167th Avenue?
    A. 4   B. 3   C. 2   D. 1

10. What fronts 1409 167th Ave?
    A. Hotel   B. Library   C. Shoestore   D. Beauty salon

11. If a ladder truck responding to this emergency were traveling South on McKinnley to get to 167th Avenue, which way would it need to turn to arrive at the scene of the fire?
    A. West   B. North   C. Southeast   D. East

MEMORY RECALL

12. The intersection of Bloomington Avenue and 167th is a
    A. Four-way stop
    B. Two-way stop
    C. Cloverleaf overpass
    D. Traffic-light-controlled intersection

13. How tall is the building fronting 1317 167th Ave?
    A. Single level
    B. Two story
    C. Three story
    D. There is no such address on 167th Avenue.

14. What building has metered parking in front?
    A. 1409 McKinnley
    B. 1413 167th Avenue
    C. 1300 Bloomington
    D. There are no parking meters in the sketch.

15. Where is the fire in relation to the hotel?
       A. Northeast   B. East   C. South   D. West

COMPLETE FIRE-FIGHTERS EXAM PREPARATION BOOK

ANSWER SHEET FOR SAMPLE SKETCH I

1. Ⓐ Ⓑ Ⓒ Ⓓ
2. Ⓐ Ⓑ Ⓒ Ⓓ
3. Ⓐ Ⓑ Ⓒ Ⓓ
4. Ⓐ Ⓑ Ⓒ Ⓓ
5. Ⓐ Ⓑ Ⓒ Ⓓ

6. Ⓐ Ⓑ Ⓒ Ⓓ
7. Ⓐ Ⓑ Ⓒ Ⓓ
8. Ⓐ Ⓑ Ⓒ Ⓓ
9. Ⓐ Ⓑ Ⓒ Ⓓ
10. Ⓐ Ⓑ Ⓒ Ⓓ

11. Ⓐ Ⓑ Ⓒ Ⓓ
12. Ⓐ Ⓑ Ⓒ Ⓓ
13. Ⓐ Ⓑ Ⓒ Ⓓ
14. Ⓐ Ⓑ Ⓒ Ⓓ
15. Ⓐ Ⓑ Ⓒ Ⓓ

ANSWERS CAN BE FOUND ON PAGE 33.

MEMORY RECALL

D. DESCRIPTIVE PASSAGES

Another form of memory exercise seen on past exams tests literal reading comprehension. You will be given an article that may pertain to an emergency situation or some technical or procedural issue.

This kind of test differs somewhat from other reading comprehension exams because in most cases you are not required to draw inferences. Rather, it involves literal comprehension (the memorization of facts as they appear in the article). Trivial items in the reading are just as important as main concepts. Try to create some sort of bizarre story as you read through the article. Associating your story with the items mentioned will help you remember the reading.

Regardless of the article's content, you will be given only five minutes to read. When your time is up for studying the article, the test examiner will collect the readings and then direct you to answer related questions in your test booklet or supplement.

SAMPLE DESCRIPTIVE PASSAGE I

Study the passage for five minutes. When your time is up, cover the passage and proceed with the questions given.

At 7:30 a.m. on March 19th, a truck transporting liquid propane (LP) gas jackknifed on an overpass, presumably because of a heavy frost on the road's surface. The driver of the truck, Sam Morton, was unhurt, and the tank's structural integrity was not compromised. However, the accident did obstruct both lanes of traffic on the overpass, causing a substantial backup of vehicles.

After 15 minutes a Washington State Patrol (WSP) officer by the name of Tom Jenkins arrived at the scene of the accident. Two additional WSP units were called to help re-route traffic, and the local fire department was notified as well. Fire station #201, located on the 1200 block of Pacific Avenue, received the dispatch because it is only six blocks from the accident. The lieutenant on duty at the time was Mitch Cosner, a ten-year veteran firefighter. The dispatch had indicated there were no injuries or fire; nevertheless, Cosner's prior training and experience dictated that it would be best to have dry chemical extinguishers on hand in the event of an LP gas fire. Fortunately, LP is a fairly stable substance to handle, and since no tank rupture or broken valve was involved, the chances of combustion occurring were fairly remote. Within seven minutes after receiving the dispatch, Cosner's fire crew arrived at the accident scene. They remained there until a heavy-duty towing rig from Paul's Towing safely removed the truck. Traffic was back to normal by 9:15 a.m.

COMPLETE FIRE-FIGHTERS EXAM PREPARATION BOOK

1. What time of day did the accident occur?
    A. 7:15 a.m.   B. 7:30 p.m.   C. 7:30 a.m.   D. 7:15 p.m.

2. What was the name of the truck driver involved in the accident?
    A. Tom Perkins   B. Sam Morton   C. Will Cosner   D. Bill Hennessy

3. How long was it before Tom Jenkins arrived at the scene of the accident?
    A. 5 minutes   B. 10 minutes   C. 15 minutes   D. 20 minutes

4. How many police units were needed to control the situation?
    A. 0   B. 1   C. 2   D. 3

5. In what state did this accident occur?
    A. It wasn't mentioned in the passage
    B. Washington, D.C.
    C. Washington
    D. Oregon

6. What was the county's numerical designation for the station that received the emergency dispatch?:
    A. 102   B. 201   C. 203   D. 101

7. What kind of extinguishing agent was mentioned as being effective against liquid propane fires?
    A. Water
    B. Foam
    C. Dry chemical
    D. None of the above

8. For how long did traffic have to be detoured before the overpass was reopened?
    A. 1 hour and 45 minutes
    B. 1 hour and 30 minutes
    C. 2 hours
    D. 2 hours and 15 minutes

9. What was Mitch Cosner's position in the fire department?
    A. Lieutenant
    B. Battalion chief
    C. Emergency medical technician
    D. Mitch Cosner doesn't work for the fire department

10. After receiving the emergency dispatch, what was the response time for firefighting personnel to arrive at the scene?
    A. 6 minutes   B. 7 minutes   C. 8 minutes   D. 9 minutes

11. What was suspected as being the primary cause of the accident?
    A. Faulty brakes.
    B. The truck driver falling asleep at the wheel.
    C. Water standing in the road causing hydroplaning.
    D. Frost on the road.

MEMORY RECALL

12. What was the name of the towing company that removed the truck?
    A. AAA   B. Cosner's   C. Paul's   D. Budget

13. What was the street location of the fire station involved?
    A. 300 block of Pacific Avenue
    B. 1200 block of Pacific Avenue
    C. 1400 block of Pacific Avenue
    D. 1200 block of 12th Street

14. Mr. Cosner had been a firefighter how many years?
    A. Mr. Cosner is not a firefighter.
    B. 10
    C. 8
    D. 5

15. Why was the fire station on Pacific Avenue dispatched to handle this particular accident?
    A. It was the only station in the county.
    B. The fire crew was specifically trained for liquid propane fires.
    C. Because Station #201 was preoccupied with another accident involving several vehicles.
    D. Because of its close proximity to the scene.

ANSWER SHEET FOR SAMPLE DESCRIPTIVE PASSAGE I

1. Ⓐ Ⓑ Ⓒ Ⓓ
2. Ⓐ Ⓑ Ⓒ Ⓓ
3. Ⓐ Ⓑ Ⓒ Ⓓ
4. Ⓐ Ⓑ Ⓒ Ⓓ
5. Ⓐ Ⓑ Ⓒ Ⓓ

6. Ⓐ Ⓑ Ⓒ Ⓓ
7. Ⓐ Ⓑ Ⓒ Ⓓ
8. Ⓐ Ⓑ Ⓒ Ⓓ
9. Ⓐ Ⓑ Ⓒ Ⓓ
10. Ⓐ Ⓑ Ⓒ Ⓓ

11. Ⓐ Ⓑ Ⓒ Ⓓ
12. Ⓐ Ⓑ Ⓒ Ⓓ
13. Ⓐ Ⓑ Ⓒ Ⓓ
14. Ⓐ Ⓑ Ⓒ Ⓓ
15. Ⓐ Ⓑ Ⓒ Ⓓ

ANSWERS CAN BE FOUND ON PAGE 33.

SAMPLE DESCRIPTIVE PASSAGE II
Study the passage for five minutes. When your time is up, cover the passage and answer the questions given.

On June 5, 1989, the Bledsoe Apartments at 14015 SE Cascade Trail came very close to being destroyed by fire. The structure had been built to code fifteen years earlier, but by today's standards the fire protection equipment installed is marginal, at best. This deficiency had been highlighted by a fire safety inspection conducted exactly one year earlier. The owner of the building, E.C.A. Property Management, an out-of-state investment partnership, gave assurances to the fire department that a new fire protection system would be installed within six months.

However, when fire inspector Roy Hollis, went back on November 30 to check on the improvements that were to have been made, none were found. A warning quickly followed, including the prospect of expensive citations if nothing were done within six months.

On the morning of June 5, Martin Jones, the apartment manager, was tired of replacing fuses for Mrs. Southerland in Apartment 103. Instead of investigating what was causing the overload, Mr. Jones replaced all of her 15-amp fuses with 35-amp fuses. Around 11 a.m., Mrs. Southerland began to smell smoke in her apartment and promptly called the manager again. By the time he arrived, the utility room was on fire and smoke was quickly filling the two-bedroom apartment. Only one of three smoke detectors in the apartment sounded an alarm, which further illustrated the inadequacy of the fire protection systems in place.

Emergency 911 was called at 11:05 a.m., and the 12 other tenants were evacuated immediately. Firefighters arrived within ten minutes and, in accordance with their pre-fire planning, set out to extinguish the blaze. Despite the fact that a truck and a four-door sedan (respective license numbers HNC 105 and LVP 306) were illegally parked in front of the fire hydrant closest to the apartment, the fire was completely extinguished by 12:15 p.m. Structural and smoke damage was confined to Apartments 102, 103, and 104. Pursuant to an investigation, the building was condemned until improvements rendered it in full compliance with current building and fire-safety codes.

1. What was the address of the apartment complex involved?
   A. 15041 E Cascade Trail
   B. 14051 SE Cascade Trail
   C. 14015 SE Cascade Trail
   D. 14051 E Cascade Trail

2. When were the Bledsoe apartments constructed?
   A. 1979   B. 1978   C. 1976   D. 1974

3. What was the date of the fire?
   A. 6/14/89   B. 6/5/89   C. 6/6/89   D. 6/30/89

4. Who was the current owner of Bledsoe Apartments?
   A. An out-of-state sole proprietorship
   B. E.A.C. Property Management
   C. E.C.A. Property Management
   D. E.A.C. Securities Division

5. What was the license number of the four-door sedan that obstructed access to the fire hydrant?
   A. LVP 306   B. HNC 105   C. LPV 360   D. HCN 150

6. How many smoke detectors failed to operate in Apartment 103?
   A. 4   B. 3   C. 2   D. 1

7. How many apartments did this complex have, excluding the manager's?
   A. 10   B. 11   C. 13   D. 15

8. How much time elapsed from the point Emergency 911 was notified until the fire was extinguished?
   A. 1 hour 5 minutes
   B. 1 hour 10 minutes
   C. 1 hour 20 minutes
   D. 1 hour 25 minutes

9. What time was it when firefighters arrived at the scene of the fire?
   A. 11:30 a.m.   B. 11:20 a.m.   C. 11:40 a.m.   D. 11:15 a.m.

10. When was the fire safety inspection that highlighted the inadequacy of fire protection equipment installed in the apartment complex conducted?
    A. 6/6/88   B. 6/6/87   C. 5/6/87   D. 6/5/88

11. What was the amperage difference between the original fuses and those that were used as replacements by the apartment manager?
    A. 10 amps   B. 15 amps   C. 20 amps   D. 35 amps

12. How many apartments suffered varying degrees of smoke and structural damage from the fire?
    A. 4   B. 3   C. 2   D. 1

13. What was the name of the fire inspector mentioned in the passage?
    A. Mr. Roy Hollis
    B. Mrs. Southerland
    C. Mr. Martin Jones
    D. None of the above

14. What time was it when the tenant in Apartment 103 started to smell smoke?
    A. 10:30 p.m.   B. 11:00 p.m.   C. 11:00 a.m.   D. 11:05 a.m.

15. In what room did the fire originate?
    A. The kitchen in Apartment 105
    B. The utility room in Apartment 103
    C. The living room in Apartment 103
    D. The guest bedroom in Apartment 104

ANSWER SHEET FOR SAMPLE DESCRIPTIVE PASSAGE II

1. Ⓐ Ⓑ Ⓒ Ⓓ
2. Ⓐ Ⓑ Ⓒ Ⓓ
3. Ⓐ Ⓑ Ⓒ Ⓓ
4. Ⓐ Ⓑ Ⓒ Ⓓ
5. Ⓐ Ⓑ Ⓒ Ⓓ

6. Ⓐ Ⓑ Ⓒ Ⓓ
7. Ⓐ Ⓑ Ⓒ Ⓓ
8. Ⓐ Ⓑ Ⓒ Ⓓ
9. Ⓐ Ⓑ Ⓒ Ⓓ
10. Ⓐ Ⓑ Ⓒ Ⓓ

11. Ⓐ Ⓑ Ⓒ Ⓓ
12. Ⓐ Ⓑ Ⓒ Ⓓ
13. Ⓐ Ⓑ Ⓒ Ⓓ
14. Ⓐ Ⓑ Ⓒ Ⓓ
15. Ⓐ Ⓑ Ⓒ Ⓓ

ANSWERS CAN BE FOUND ON PAGE 33.

ANSWER SHEET FOR MEMORY RECALL SAMPLE QUESTIONS

SAMPLE FLOOR PLAN I

1. C	6. D	11. C
2. C	7. D	12. B
3. B	8. A	13. C
4. D	9. D	14. D
5. A	10. B	15. A

SAMPLE FLOOR PLAN II

1. C	6. D	11. C
2. D	7. C	12. A
3. B	8. A	13. C
4. A	9. D	14. D
5. B	10. B	15. D

SKETCH I

1. D	6. C	11. A
2. C	7. D	12. D
3. A	8. B	13. C
4. A	9. A	14. D
5. B	10. D	15. B

DESCRIPTIVE PASSAGE I

1. C	6. B	11. D
2. B	7. C	12. C
3. C	8. A	13. B
4. D	9. A	14. B
5. C	10. B	15. D

DESCRIPTIVE PASSAGE II

1. C	6. C	11. C
2. D	7. C	12. B
3. B	8. B	13. A
4. C	9. D	14. C
5. A	10. D	15. B

Your score for each exercise would rate as follows:

14–15 correct	EXCELLENT
12–13 correct	GOOD
10–11 correct	FAIR
less than 10 correct	POOR

MECHANICAL APTITUDE

Firefighters have a wide array of tools and equipment at their disposal. If a firefighter understands each implement and its implicit function, his or her job can be significantly simplified. It is also important to understand how mechanized parts such as gears, pulleys and belts, hydraulics, etc., transmit energy to perform work.

This does not mean you need a degree in mechanical engineering, or must seek an apprenticeship with a host of professional contractors to be a firefighter. However, it is fair to expect a firefighter applicant to have a general sense of the tools used in carpentry, electrical work, plumbing, and auto repair. Such information can be acquired through vocational courses, or by various repairing or building projects. You will also be expected to have a grasp of general mechanical principles concerning leverage, mechanical advantage, and fluid dynamics. Questions requiring knowledge of specialized firefighting equipment and application will not be on the exam.

For study, this chapter will be broken into two parts. The first section will be a general overview of tools classified by trade use. A sampling of questions will follow, complete with answers and explanations. The second section will be a general review of mechanical principles and their applications.

A. TOOLS
I. TOOLS THAT DETERMINE MEASUREMENT

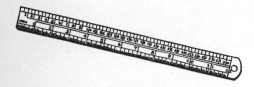

STEEL RULER
Has linear gradations (metric or English) to measure a workpiece; can double as a straightedge. (Average length is 12 inches).

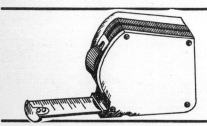

STEEL TAPE MEASURE (Push-pull tape measure)
Has linear gradations (metric or English) on a flexible metal tape. The tape automatically retracts into a small housing for convenience. (Normal length is 3 to 16 feet.)

WIND-UP TAPE MEASURE
Has linear gradation (metric or English) on a flexible metallic tape used to measure. The tape can be retracted into a circular housing by the use of a small crank on one side of the housing. (Normal length is 30 to 100 feet.)

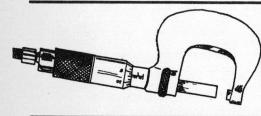

MICROMETER CALIPERS
Used in determining precise measurements. Calipers can measure from 0 to 12 inches or 0 to 300 millimeters, depending on the type selected.

MECHANICAL APTITUDE

VERNIER CALIPERS
Used in determining precise internal or external measurements. Measuring capacity is 6 to 72 inches or 150 to 1800 millimeters.

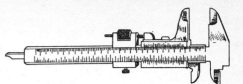

SLIDE CALIPERS
Use is comparable to the vernier calipers; however, the gradations are not as precise. Its measuring capacity is up to 3 inches.

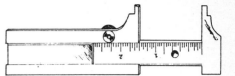

OUTSIDE CALIPERS (Bow calipers)
Used to transfer external measurements from a rule, or to match two items to fit. Measuring capacity is up to 12 inches.

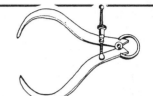

INSIDE CALIPERS
Used to transfer internal measurements from a ruler, or to match two items to fit.

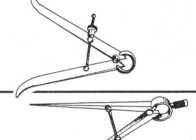

SPRING DIVIDERS (Bow compass)
Used to inscribe circles or arcs or to calibrate equal divisions on a line.

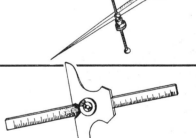

DEPTH GAUGE
Used to determine depths of holes, grooves, and mortices.

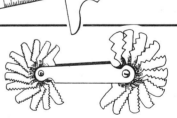

SCREW PITCH GAUGE
Used to determine the pitch of a machined thread. Each blade is calibrated in metric units.

II. TOOLS THAT DETERMINE ANGULATION

CARPENTER'S STEEL SQUARE (Framing square)
Used to check the squareness of framing (i.e., 90-degree or right angles). There are metric or English-system gradations for measuring as well.

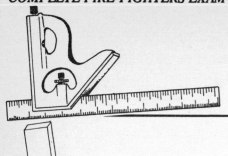

COMBINATION SQUARE
An all-purpose tool that can measure and serve as a try square, miter square, and level. Measuring capacity in metric or English gradations is 12 inches.

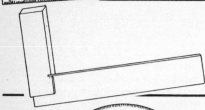

TRY SQUARE
Used to determine if a workpiece is square (i.e., has 90-degree right angles).

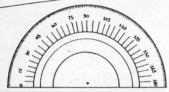

PROTRACTOR
Used as an instrument to draw or plot angles (0 to 180 degrees) on a surface.

T-BEVEL
Used to mark or verify angles on a workpiece. Normally a protractor is used to serve as a point of reference.

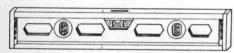

LEVEL
Used to determine a true horizontal line or accurate level of a surface.

PLUMB BOB
Used to determine a true vertical line.

III. TOOLS AS CLASSIFIED BY USE

1. SAWS

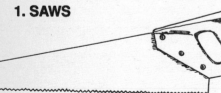

HANDSAW (Crosscut, rip, or combination)
Used to cut wood planks, sheets, or panels to desired size.

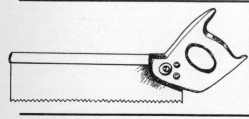

BACK SAW (Tenon saw)
Primarily used to cut fine joints in wood, as in a tenon, lap, or dovetail. It is can also be used in conjunction with a miter box to cut accurate 45- or 90-degree angles.

MECHANICAL APTITUDE

COPING SAW
Used in making curved cuts in wood or plastic.

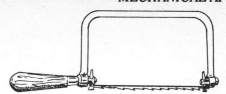

COMPASS SAW (Keyhole saw)
Used to cut holes in a panel. This saw is not restricted to an edge of a workpiece, like the coping saw, because of the lack of a frame.

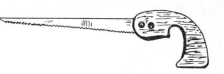

HACKSAW
Used to cut metal sheets, pipes, plastics, etc.

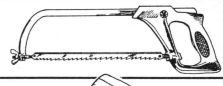

PORTABLE CIRCULAR SAW
Used to cut lumber or plywood to desired size. Saw blades that can be used with this tool are:

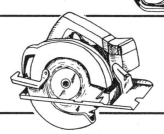

1. Cross cut—cuts across grain of lumber
2. Rip cut—cuts parallel to the grain of lumber
3. Combination—cuts lumber in any direction
4. Carbide tipped—cuts particle board or lumber with the primary advantage of teeth remaining sharper for longer periods of time.
5. Metal cutting—cuts thin sheets of copper, aluminum, and other metals.
6. Friction blade—cuts corrugated sheet metal
7. Abrasive disk—cuts concrete, marble, etc.

RECIPROCATING SAW
Used in much the same way as a hand or compass saw to cut wood, plastic, or thin-gauge metal depending on blade selection.

SABER SAW (Powered jigsaw)
Used to cut curves in various materials depending on blade selection.

CHAIN SAW
Used to fell timber and cut logs to desired lengths.

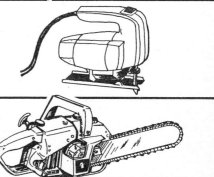

2. HAMMERS

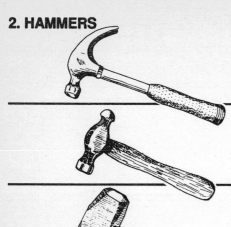

CLAW HAMMER
Used for general carpentry and can pull and extract nails.

BALL PEEN HAMMER (Engineer's hammer):
Used in metalworking such as rivet setting or metal forging.

CLUB HAMMER (Hand-drilling hammer)
Used for heavy-duty work involving masonry drills or chisels. The head can weigh up to four pounds.

SLEDGE HAMMER
Used for heavy-duty work such as breaking concrete, driving posts, etc. The head can weigh from two to twenty pounds.

SOFT-FACED MALLET (Rubber mallet)
Used to drive materials without causing any damage to the tools involved or to the surface of a workpiece.

3. SCREWDRIVERS

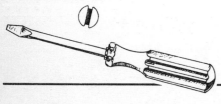

CONVENTIONAL STRAIGHT-SHANK SCREWDRIVER
Used to drive slotted head screws.

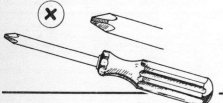

PHILLIPS SCREWDRIVER
Used to drive Phillips-head screws. Its design enhances one's grip compared to that achieved with conventional slotted- or flared-tip screwdrivers.

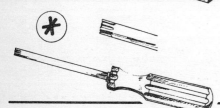

TORX SCREWDRIVER
Used when a fair amount of torque is required on a screw. This design enhances your grip still more than a Phillips or flared-tipped screwdriver.

MECHANICAL APTITUDE

STUBBY SCREWDRIVER (Close-quarter screwdriver)
Used to drive screws where space is fairly restricted.

JEWELER'S SCREWDRIVER
Used in driving or loosening relatively small screws such as those found in eyeglass hinges, clocks, wrist watches, etc.

OFFSET SCREWDRIVER (Cranked screwdriver)
Used when standard screwdrivers can not fit. Additionally, these screwdrivers enable the user to place a fair amount of torque on the screw being driven.

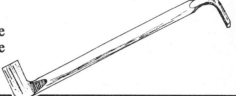

4. WRENCHES

OPEN-ENDED WRENCH
Used to tighten or loosen nuts and bolts from the side. This is advantageous over other wrenches in cramped areas where a nut is obstructed by something directly overhead.

BOX WRENCH
Used to tighten or loosen nuts and bolts. A closed wrench of this design allows for significantly more torque to be applied without the fear of the wrench slipping and stripping the nut. Its drawback is that it takes more time to use this wrench than the open variety.

OFFSET WRENCH
Designed to be able to reach a nut or bolt in a recessed area over an obstruction. These wrenches also have the advantage of allowing hand clearance when tightening a bolt or nut flush with a work surface.

CRESCENT WRENCH
An adjustable wrench that can be used to tighten or loosen nuts and bolts of varied sizes.

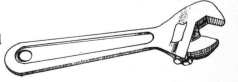

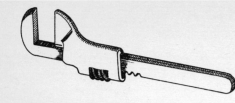
MONKEY WRENCH (Screw wrench)
Another adjustable wrench that can be used to tighten or loosen nuts and bolts of varied sizes. This wrench has a heavier design than a crescent wrench.

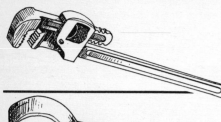
STILSON WRENCH (Pipe wrench)
An adjustable wrench used to grip round objects like metal pipes or steel rods.

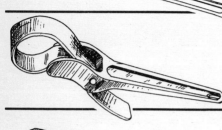
STRAP WRENCH
Used for gripping and turning pipes without harming the exterior finish.

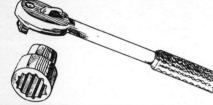
SOCKET WRENCH
One of the most common wrenches used today. Nuts and bolts can be quickly tightened or loosened with the ratchet handle, thus sparing the need to readjust for another grip as with most other wrenches.

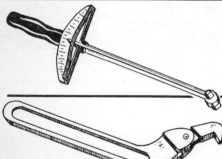
TORQUE WRENCH
Used to apply a calibrated force when tightening a nut or bolt as recommended by the manufacturer.

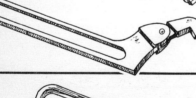
SPANNER WRENCH (Hook spanner)
Used to turn special nuts or hose couplings.

ALLEN WRENCH (Setscrew wrench)
An L-shaped hexagonal key used to tighten or loosen a machined setscrew.

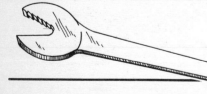
CROCODILE WRENCH (Bulldog wrench)
Used to grip and turn round stock such as pipe or steel rod.

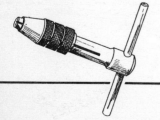
TAP WRENCH
Used when cutting internal screw threads with a tap. It allows for a better and more even grip when applying pressure.

MECHANICAL APTITUDE

5. PLIERS

NEEDLE-NOSE PLIERS (Snipe-nosed pliers)
Used to shape or cut thin-strand wire and grip small items (e.g., washers, nuts, etc.) in confined spaces.

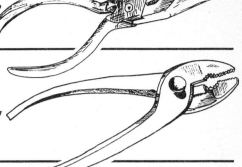

SLIP JOINT PLIERS
Used like ordinary pliers, but a pivot allows for two different jaw settings.

CURVED JAW PLIERS (Channel lock pliers)
Used much in the same manners as slip joint pliers. ere are additional settings to adjust jaw widths to accommodate larger round stock.

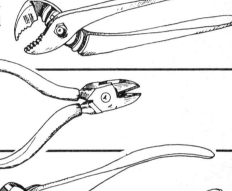

SIDE-CUTTING PLIERS
Used to cut metal wire.

END-CUTTING PLIERS
Used to crop metal wire close to a work surface.

---

6. POWER SANDERS

BELT SANDER:
Used to quickly sand wood, steel, or plastic. Various abrasives embedded in the belts are used on this machine in a back and forth motion.

DISK SANDER
Used for the same purpose as a belt sander, but the abrasives are embedded in disks that are used in conjunction with a drill or offset sander grinder.

ORBITAL SANDER (Finishing sander)
Used to finish sanding with light abrasive paper.

7. FILES

FLAT FILES
Used to file flat surfaces; can be used on most types of material.

TRIANGULAR FILE
Used to file angles or square corners (e.g., saw teeth).

ROUND FILE (Rat-tail file)
Used to smooth round openings.

8. DRILLS AND ACCESSORIES

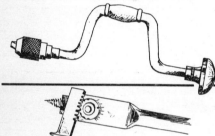

BRACE
Used to manually drill holes in wood or provide extra torque in driving screws. Bits that can be used with this tool are

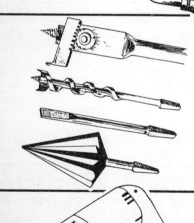

1. Expansive bit—an adjustable spurred cutter, which allows holes of various sizes to be drilled in wood.
2. Auger bit (twist bit)—various sizes of helical twists and a prominent lead screw are characteristic of this form of bit for drilling holes in wood.
3. Turn screw bit—used to drive screws.
4. Ream—used only to bevel or widen the opening of a pipe.

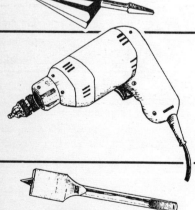

POWER DRILL
Used in principally the same way as a brace, the the difference being in speed and flexibility of using various accessories (e.g., sanding disks, buffs, wire brushes, lathe, etc.). Bits that can be used with this tool are:

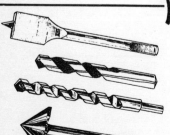

1. Spade bit (Flat bit)—used to drill wide holes in wood or plastics.
2. Twist drill (Morse drill)—used to drill holes in wood, metal or plastics.
3. Masonry drill—a carbide tip allows this bit to be used to drill holes in material such as concrete, marble, stone, etc.
4. Countersink bit—used to create a recess hole to accommodate the head of a countersink screw.

MECHANICAL APTITUDE

CENTER PUNCH
Used to mark and guide the placement of a drill point.

IV. TOOLS NORMALLY SEEN IN VARIOUS TRADES
1. PLUMBING

DRAIN SNAKE (Drain auger)
Used to clear or remove debris that obstructs a drain pipe from a sink or toilet by extending into the piping. Some drain augers can reach through up to 100 feet of pipe.

PLUNGER
Used to clear debris blocking a drain trap by creating back pressure on the obstruction.

PIPE CUTTER (Wheel cutter)
Used to cut metal pipe to the desired size.

PROPANE TORCH
Used as a heat source for brazing and soldering pipework.

SOLDERING IRON
Used to heat metal and solder to form a joint.

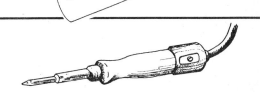

VISE GRIPS (Plier wrench)
Used to grip round metal stock or sheet metal firmly without much effort.

g. OTHER TOOLS applicable to plumbing that have already been elaborated on in SECTION III are:

Ball peen hammer	Strap wrench
Claw hammer	Crocodile wrench
Open-end wrench	Needle-nose pliers
Crescent wrench	Slip joint pliers
Monkey wrench	Channel lock pliers
Stilson wrench	Flat file
Drill	Round file

2. ELECTRICAL

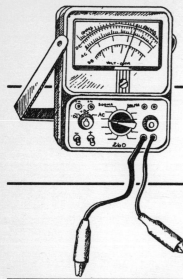

OHMMETER
Used to measure the amount of resistance in a given unit or circuit (calibrated in ohms).

AMMETER
Used to measure the amount of electrical current flow (calibrated in amperes).

VOLTMETER
Used to measure the flow of electricity through a conductor in relation to electrolytic decomposition (calibrated in volts).

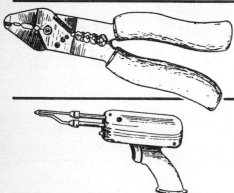

MULTIPURPOSE TOOL (Wire-stripping pliers)
Used as a pair of pliers to cut and strip insulation from wire, to crimp terminals, and to cut small screws.

SOLDERING GUN
Used to solder wire to form contact terminals and in other related light electrical work.

TEST LIGHT (Circuit tester)
Used to determine the presence of current in a given circuit.

OTHER TOOLS applicable to electrical work that have already been elaborated on in Section III are:

Crosscut handsaw	Stubby screwdriver
Compass saw	Drill
Portable circular saw	Needle-nose pliers
Claw hammer	Slip joint pliers
Ball peen hammer	Curved jaw pliers
Conventional straight-shank screwdriver	Side-cutting pliers
	End-cutting pliers

MECHANICAL APTITUDE

3. CARPENTRY

PORTABLE ELECTRIC ROUTER
Used to cut various grooves and moldings (e.g., hinge mortise, dovetail, rabbit, etc.).

JACK PLANE (fore plane)
Used to dimension or smooth a wooden workpiece.

PIPE CLAMPS
Used to secure boards or framing together while they are bonded by glue.

PARALLEL CLAMPS (handscrew)
Used principally like pipe clamps with the added feature of being able to secure an angled object, whether wood or metal.

C-CLAMPS
Used to clamp wood or metal for various purposes.

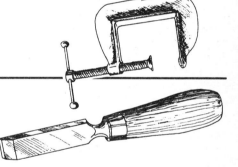

WOOD CHISELS
Used to trim or groove wood. Normally used with a soft-faced mallet. Wood chisels come in a variety of forms and sizes.

NAIL PUNCH (Nail set)
Used as an instrument to drive nails below the surface of a wooden workpiece.

AWL
Used to start holes in wood to accommodate nails or screws.

OTHER TOOLS applicable to carpentry that have already been elaborated on in SECTION III are:

Handsaw	Soft-faced mallet
Back saw	Conventional screwdriver
Coping saw	Phillips screwdriver

COMPLETE FIRE-FIGHTERS EXAM PREPARATION BOOK

Compass saw   Power sander
Circular saw   Drills
Saber saw   Slip joint pliers
Reciprocating saw   Claw hammer

4. METAL WORKING

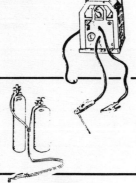

ARC WELDER
Used to fuse pieces of metal together using electrical current (rated in amperage).

CUTTING TORCH (Acetylene torch)
Used to cut metal or thick steel by means of a flame generated from an oxygen and acetylene mix.

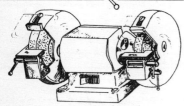

MACHINIST VISE
Used to secure or hold a metal workpiece

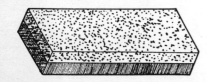

BENCH GRINDER
Used to sharpen tools or remove rough edging from a metal workpiece. Optional wire brush or buffer wheel can further clean and polish metal work.

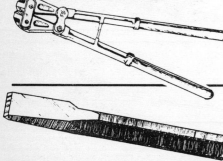

WHETSTONE
A rectangular stone comprised of gritty abrasive that is used to sharpen tools manually.

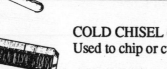

BOLT CUTTERS
Used to cut steel rod, bolts, chains, and locks through the use of compound leverage.

COLD CHISEL (Flat chisel)
Used to chip or cut cold metal such as rivets, chain links, bolts, etc.

STRAIGHT SNIPS (Tin snips)
Used to cut thin sheet metal.

SHEET METAL PUNCH
Used to make holes in sheet metal to accommodate screw nails.

ROUND SPLIT DIE IN HOLDER
Used to create threading on smooth metal rods. Comparable to manufacturing a bolt.

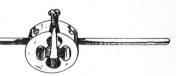

OTHER TOOLS applicable to metal working that have already been elaborated on in SECTION III are:
- Hacksaw
- Metal cutting blade for portable circular saw
- Friction blade for portable circular saw
- Saber saw with metal-cutting blade
- Ball peen hammer
- Drill

---

5. AUTOMOTIVE MECHANICS

HYDROMETER
A float device for determining the specific gravity of the electrolytes in a battery. This determines the state of charge.

GREASE GUN
Used to inject lubricant into special fittings such as those found in ball joints, steering linkage, universal joints, etc.

FILTER WRENCH (Strap wrench)
Used specifically to remove or tighten oil filters.

SPARK PLUG WRENCH
Used specifically to remove or tighten spark plugs without damaging the insulator.

FEELERS GAUGE
Used to determine gaps between electrical contacts such as those found in a spark plug. It can also measure the gap between a shaft and bearing on an engine.

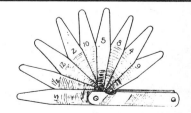

COMPLETE FIRE-FIGHTERS EXAM PREPARATION BOOK

TIRE GAUGE
Used to determine the proper inflation pressure in a tire.

COMPRESSION GAUGE
Used to measure the amount of compression an engine cylinder possesses.

6. MISCELLANEOUS TOOLS

GLASS CUTTER
Used to score a line across glass. Applying pressure by bending the glass then causes it to break along the scored line. The notches on the back of the cutter are designed to remove small pieces of glass from the desired cut or line.

UTILITY KNIFE
Used as a general-purpose knife to cut various materials.

WRECKING BAR (Crowbar)
Used as a lever to pry things apart or remove nails

CAULKING GUN
Used to apply sealants to various joints such as those found in windows, door frames, roofing vents, and bathroom tile.

STAPLE GUN
Used to drive staples in attaching various materials together.

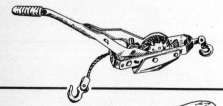

COME-ALONG WINCH
Used to exert a pulling force (vertical or horizontal) on heavy objects (not exceeding five tons).

MASON'S TROWEL (Brick trowel)
Used to spread, shape, and smooth mortar when working with bricks or concrete block.

MECHANICAL APTITUDE

SKIMMER FLOAT
Used to acquire a smooth finish when working with either wet plaster or concrete.

LOPPING SHEARS
Used to prune back branches of shrubs or trees.

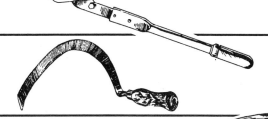

SICKLE
Used to cut tall weeds or grass.

WEDGE FELLING AXE
A cutting tool used for felling trees and chopping or hewing wood.

HATCHET
Used to trim wood and doubles as a hammer.

PICK AXE
Used to indent, pierce, or break up hard material such as concrete, asphalt, hardpan soils, etc.

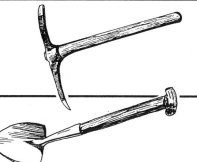

SHOVEL
Used to lift and throw earth or other materials.

SAMPLE QUESTIONS ABOUT TOOLS AND THEIR USE

1. Which of the following tools would be used to get precise metric dimensions of a pipe's interior diameter?
   A. Micrometer calipers
   B. Vernier calipers
   C. Depth gauge
   D. Push-pull tape measure

2. Which of the following tools would best be used to determine the gauge or thickness of a thin sheet of metal?
   A. Carpenter's square
   B. Depth gauge
   C. Inside calipers
   D. Micrometer calipers

3. A combination square is versatile because it can be used
   A. As a miter square
   B. To measure
   C. To level
   D. All of the above

4. The diagram shown is that of a

   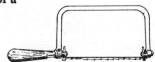

   A. Miter saw
   B. Crosscut handsaw
   C. Coping saw
   D. Portable circular saw

5. What kind of blade would be recommended to cut a plank of wood lengthwise (with the grain)?
   A. Rip cut   B. Cross cut   C. Dado blade   D. Abrasive disk

6. A wood chisel is best used with what kind of hammer?
   A. Claw hammer
   B. Ball peen hammer
   C. Sledge hammer
   D. Rubber mallet

7. What kind of tool would have to be used to tighten the screw shown in the diagram?

   A. Torx screwdriver
   B. Allen wrench
   C. Phillips screwdriver
   D. Socket wrench

MECHANICAL APTITUDE

8. If a painted metal pipe needed to be turned to thread it into a fixed coupling, what kind of wrench would probably be most suitable?
    A. Stilson wrench
    B. Torque wrench
    C. Strap wrench
    D. Crocodile wrench

9. What kind of tool would probably be used to eliminate internal burrs or rough edging of a pipe that has just been cut?
    A. Ream
    B. Flat file
    C. Orbital sander
    D. Expansive drill bit and brace

10. What kind of drill bit would be necessary to drill a hole into concrete flooring?
    A. Spade bit   B. Auger bit   C. Ream   D. Masonry bit

11. Which of the following wrenches is not adjustable?
    A. Crescent wrench
    B. Crocodile wrench
    C. Monkey wrench
    D. Stilson wrench

12. If two angled pieces of wood were to be glued together as shown in the diagram, what kind of clamps would probably be used?

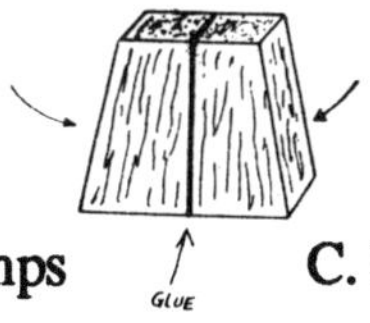

    A. Parallel clamps   B. C-clamps   C. Machinist vise   D. Pipe clamps

13. The tool portrayed in the diagram is a/an?

    A. Awl   B. Sheet metal punch   C. Center punch   D. Nail punch

14. If a piece of ½-inch rebar had to be shortened, what kind of a saw would you use?
    A. Portable circular saw with a combination blade
    B. Back saw
    C. Coping saw
    D. Hacksaw

15. Prior to staining or painting furniture, what kind of power sander would be used to do the smooth finish work?
    A. Belt sander   B. Disk sander   C. Orbital sander   D. All of the above

16. A hydrometer measures what?
    A. Glycerol in an automotive coolant system
    B. Battery electrolyte fluid (wet cell)
    C. Hydraulic fluid levels
    D. None of the above

17. What tool is practically a trademark for the electrical profession?
    A. Multipurpose tool   B. Plunger   C. Claw hammer   D. Slip joint pliers

18. Which of the following wrenches are the slowest to use in loosening a nut?
    A. Crescent wrench
    B. Open-ended wrench
    C. Screw wrench
    D. Box wrench

19. A jack plane is associated with what profession?
    A. Plumbing   B. Electrical   C. Carpentry   D. Metal fabrication

20. What kind of wrench could loosen the nut shown in the diagram?

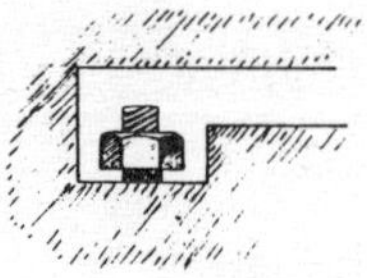

    A. Crescent wrench
    B. Open-ended wrench
    C. Offset box wrench
    D. Monkey wrench

21. What kind of pliers are shown in the diagram?

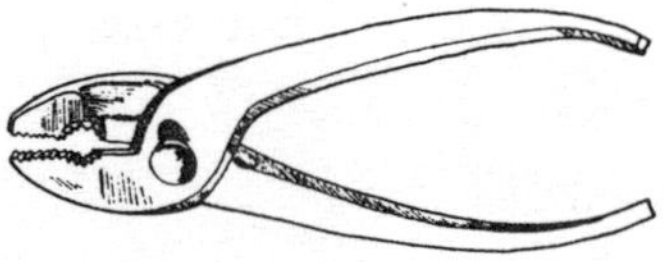

    A. Channel lock pliers
    B. Slip joint pliers
    C. Needle-nose pliers
    D. End-cutting pliers

22. If the contents of a wooden shipping crate needed to be inspected by custom officials, what one tool would probably be used to open it?
    A. Wrecking bar   B. Wood chisel   C. Sledge hammer   D. Chain saw

23. Which of the following gauges are used to determine spark plug gaps?
    A. Screw pitch gauge
    B. Depth gauge
    C. Compression gauge
    D. Feelers gauge

24. If a rivet had to be removed from sheet metal, what tool could be used?
    A. Bolt cutters   B. Lopping shears   C. Hacksaw   D. Cold chisel

25. When a plumber is said to be sweating a joint, what is the probable source of heat?
    A. Arc welder
    B. Oxyacetylene cutting torch
    C. Propane torch
    D. All of the above

MECHANICAL APTITUDE

26. If a lawn mower blade had to be sharpened, what tool would be the quickest and most convenient to use?
    A. Bench grinder   B. Flat file   C. Whetstone   D. Emery paper

27. The diagram shown illustrates what kind of a tool?

    A. Inside calipers   B. Outside calipers   C. Bow compass   D. Micrometer

28. A reciprocating saw can be used to cut through which of the following materials?
    A. Plastics   B. Particle board   C. Dry wall   D. All of the above

29. If a contractor needed a true vertical line of reference, what tool would be used?
    A. Carpenter's square   B. Bevel square   C. Plumb bob   D. Level

30. What kind of handsaw is used specifically to cut tenon joints?
    A. Back saw   B. Coping saw   C. Hacksaw   D. Saber saw

31. What kind of screwdriver would be needed to drive the screw shown in the diagram?

    A. Conventional straight shank
    B. Phillips
    C. Torx
    D. Close-quarters screwdriver

32. What is the reason a spark plug wrench should be used to remove spark plugs over other type of wrenches?
    A. It's quicker
    B. More torque can be applied
    C. It lessens the chance of getting burned
    D. Prevents potential damage from occurring to the insulator or electrode

33. What is shown in the diagram?

    A. Utility knife   B. Glass cutter   C. Awl   D. Triangular file

34. Which of the following drill bits is not used with a brace?
    A. Auger bit   B. Expansive bit   C. Twist bit   D. Masonry drill bit

35. What pair of pliers can apply a constant amount of pressure without much effort?
    A. Needle-nose pliers
    B. Curved jaw pliers
    C. Plier wrench
    D. Slip joint pliers

36. What is the measurement as shown by the micrometer caliper in the diagram?

    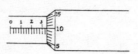

    A. 0.320 inches   B. 0.335 inches   C. 0.35 inches   D. 0.275 inches

37. What is the name of the tool shown in the diagram?

    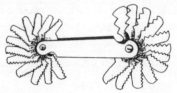

    A. Screw pitch gauge   B. Wire gauge   C. Depth gauge   D. Feelers gauge

38. If an ammeter registers 0, what does that tell you?
    A. There is too much resistance to complete a circuit.
    B. There must be a short somewhere.
    C. The wire in question is considered hot.
    D. No electrical current is flowing.

39. What is the primary advantage of a carbide-tipped saw blade?
    A. It is sharper.
    B. It can cut quicker with a smaller chance of kick-back.
    C. The hardened teeth will remain sharper longer.
    D. It can cut metal and concrete.

40. What advantage does an offset screwdriver have over the other screwdrivers?
    A. It can tighten or loosen screws in cramped places.
    B. It is quicker to use in tightening screws as compared to conventional shank screwdrivers.
    C. It can exert substantially more torque than regular screwdrivers.
    D. Both A and C are correct.

41. What kind of a tool is used to cut external screw threads on round metal stock?
    A. Taps in a tap wrench
    B. Round split dies in a holder
    C. Pipe cutter
    D. Stilson wrench

42. What tool would be most appropriate to use in cutting thin sheet metal such as tin?
    A. Cold chisel
    B. Straight snips
    C. Side-cutting pliers
    D. End-cutting pliers

43. What is the name of the tool shown in the diagram?

    A. Torque wrench   B. Crocodile wrench   C. Stilson wrench   D. Monkey wrench

44. If you wanted to cut down a tree using hand tools, you would probably use a
    A. Pickaxe   B. Hatchet   C. Sickle   D. Wedge felling axe

45. The tool pictured is associated with what trade?

    A. Plumbing
    B. Electrical work
    C. Masonry
    D. Automotive mechanic

ANSWER SHEET FOR SAMPLE TOOL QUESTIONS

1. Ⓐ Ⓑ Ⓒ Ⓓ
2. Ⓐ Ⓑ Ⓒ Ⓓ
3. Ⓐ Ⓑ Ⓒ Ⓓ
4. Ⓐ Ⓑ Ⓒ Ⓓ
5. Ⓐ Ⓑ Ⓒ Ⓓ
6. Ⓐ Ⓑ Ⓒ Ⓓ
7. Ⓐ Ⓑ Ⓒ Ⓓ
8. Ⓐ Ⓑ Ⓒ Ⓓ
9. Ⓐ Ⓑ Ⓒ Ⓓ
10. Ⓐ Ⓑ Ⓒ Ⓓ
11. Ⓐ Ⓑ Ⓒ Ⓓ
12. Ⓐ Ⓑ Ⓒ Ⓓ
13. Ⓐ Ⓑ Ⓒ Ⓓ
14. Ⓐ Ⓑ Ⓒ Ⓓ
15. Ⓐ Ⓑ Ⓒ Ⓓ

16. Ⓐ Ⓑ Ⓒ Ⓓ
17. Ⓐ Ⓑ Ⓒ Ⓓ
18. Ⓐ Ⓑ Ⓒ Ⓓ
19. Ⓐ Ⓑ Ⓒ Ⓓ
20. Ⓐ Ⓑ Ⓒ Ⓓ
21. Ⓐ Ⓑ Ⓒ Ⓓ
22. Ⓐ Ⓑ Ⓒ Ⓓ
23. Ⓐ Ⓑ Ⓒ Ⓓ
24. Ⓐ Ⓑ Ⓒ Ⓓ
25. Ⓐ Ⓑ Ⓒ Ⓓ
26. Ⓐ Ⓑ Ⓒ Ⓓ
27. Ⓐ Ⓑ Ⓒ Ⓓ
28. Ⓐ Ⓑ Ⓒ Ⓓ
29. Ⓐ Ⓑ Ⓒ Ⓓ
30. Ⓐ Ⓑ Ⓒ Ⓓ

31. Ⓐ Ⓑ Ⓒ Ⓓ
32. Ⓐ Ⓑ Ⓒ Ⓓ
33. Ⓐ Ⓑ Ⓒ Ⓓ
34. Ⓐ Ⓑ Ⓒ Ⓓ
35. Ⓐ Ⓑ Ⓒ Ⓓ
36. Ⓐ Ⓑ Ⓒ Ⓓ
37. Ⓐ Ⓑ Ⓒ Ⓓ
38. Ⓐ Ⓑ Ⓒ Ⓓ
39. Ⓐ Ⓑ Ⓒ Ⓓ
40. Ⓐ Ⓑ Ⓒ Ⓓ
41. Ⓐ Ⓑ Ⓒ Ⓓ
42. Ⓐ Ⓑ Ⓒ Ⓓ
43. Ⓐ Ⓑ Ⓒ Ⓓ
44. Ⓐ Ⓑ Ⓒ Ⓓ
45. Ⓐ Ⓑ Ⓒ Ⓓ

ANSWERS TO SAMPLE TOOL QUESTIONS

1. **B.** Vernier calipers have calibrated internal jaws that can precisely measure the inside diameter of pipe.

2. **D.** Micrometer calipers are used to determine very fine measurements of flat workpieces.

3. **D.** All of the above. A combination square can act as a miter square or try square, measure, and level.

4. **C.** Coping saw.

5. **A.** The rip cut blade is primarily used to cut lumber parallel to the grain. A dado blade is primarily used to cut grooves for various forms of joints.

6. **D.** Rubber mallet is less prone to damage the wood chisel's handle.

7. **B.** Allen wrench—an allen set screw is set apart from others by its characteristic hexagonal slot, thus requiring an allen wrench for tightening or loosening.

8. **C.** Strap wrench is designed not to harm the exterior finish of the pipe. The stilson and crocodile wrenches would probably leave tooth marks and chip the paint.

9. **A.** Reams are used for this purpose as well as beveling. A flat file is more appropriate for flatter kinds of stock instead of pipe.

10. **D.** Masonry bits are designed with carbide tips that allow penetration of hard materials like stone, marble, and concrete. Reams are not meant for drilling holes.

11. **B.** Crocodile wrench is the only wrench listed without an adjustable feature.

12. **A.** Parallel clamps have the versatility to secure or clamp angled woodwork. A machinist's vise is not meant for wood.

13. **C.** Center punch.

14. **D.** Hacksaw is the only alternative given that can cut metal.

15. **C.** Orbital sander is primarily used in finishing wood products. The other sanders are meant for coarser work and do not leave the same kind of smooth finish.

16. **B.** Battery electrolyte fluid. A hydrometer can determine if a car battery has sufficient charge or warrants replacing.

17. **A.** Multi-purpose tool. Wire-stripping pliers are a must for an electrician.

18. **D.** Box wrench is slowest because it requires the wrench to be lifted off the nut, repositioned, and replaced on the nut before it can be turned. All other wrenches mentioned as alternatives are open faced and allow for quicker turning.

MECHANICAL APTITUDE

19. *C.* Carpentry. A jack plane smooths and pares down lumber.

20. *C.* Offset box wrench. The nut is recessed with an overhead obstruction that effectively eliminates the use of the other straight shanked wrenches mentioned.

21. *B.* Slip joint pliers.

22. *A.* Wrecking bar. The alternatives could potentially damage the contents.

23. *D.* Feelers gauge.

24. *D.* Cold chisel. The rivet, being practically flush with the metal surface, does not present enough workable area for either bolt cutters or a hacksaw to work effectively. Lopping shears are used only for gardening.

25. *C.* Propane torch. Blow torches or propane torches are commonly used in plumbing to heat pipes. Arc welders and oxyacetylene cutting torches are meant to fuse and cut metal, respectively.

26. *A.* Bench grinder. The three alternatives could potentially sharpen a lawn mower blade; however, their capabilities can be used better elsewhere.

27. *B.* Outside calipers.

28. *D.* Reciprocating saw can cut through sheet metal, wood, and a host of other materials providing the proper blade is selected.

29. *C.* Plumb bob determines a true vertical reference line while a level is used for horizontal reference. Carpenters and bevel squares are used to figure angulation.

30. *A.* Back saws have fine rip saw teeth that enable them to cut a variety of joints. Saber saws are not hand saws.

31. *B.* Phillips.

32. *D.* Prevents potential damage from occurring to the insulator or electrode.

33. *B.* Glass cutter.

34. *D.* Masonry drill bits are normally used in power drills.

35. *C.* Plier wrench is another name for vise grips; can actually lock onto a piece of work with a minimum of effort. The other pliers exert pressure only in direct proportion to how hard they are squeezed. Vise grips utilize adjustable leverage to enhance a grip over and above what can be done conventionally.

36. *B.* 0.335 inches. Add the sleeve reading, 0.325 inches, and the thimble reading, 0.010.

37. *A.* Screw pitch gauge.

38. D. There is no electrical current flow. The other statements are false.

39. C. The hardened teeth remain sharper longer.

40. D. Both A and C are correct. Offset screwdrivers can work in tight places and apply more torque because of their shape.

41. B. Round split die in a holder. Taps with a tap wrench are used to cut interior screw threads in a drilled hole.

42. B. Straight snips, better known as tin snips. Side cutting and end cutting pliers are used primarily to cut wire.

43. A. Torque wrench. Such wrenches can accurately determine the amount of torque that is being applied to a given nut or bolt (e.g., cylinder-head bolts on an engine, as recommended by the manufacturer).

44. D. Wedge felling axe. A hatchet has a smaller head designed more to trim wood than to fell trees.

45. C. Masonry. A skimmer float is the tool pictured. It is used to create a smooth finish on wet concrete or plaster.

B. MECHANICAL PRINCIPLES

Humankind learned early that in order to create more force to perform a certain job, one of two things was needed: either more workers or the use of simple machines. Machines can basically be classified in six general groups; lever, pulley, inclined plane, wheel and axle, screw, and wedge. Regardless of how complicated a given machine is, one or a combination of these principles is always employed. Let's examine each.

I. SIMPLE MACHINES

1. LEVERS

For purposes of illustration, think of a plank of wood balanced on a pivot point, such as a teeter totter. (For future reference, the pivot point is called the *fulcrum*). If two different weights were placed on the plank at equal distances from the fulcrum, the side bearing the heavier weight would fall to touch the ground, lifting the lighter weight. However, if the heavier weight were moved closer to the fulcrum, eventually a balance would be struck. If it was moved even closer to the fulcrum, the lighter weight could actually have enough force to lift the heavier weight. This concept is known as *leverage* or *mechanical advantage*. This is why one person can move large rocks with a pry bar, carry heavy loads using a wheelbarrow, or apply tremendous pressure with the jaws of vise grips. To demonstrate the benefits of leverage quantitatively, the following equation can be applied:

$$\text{EFFORT} \times \text{EFFORT DISTANCE} = \text{RESISTANCE} \times \text{RESISTANCE DISTANCE}$$

Let's apply this to the following example

EXAMPLE 1:

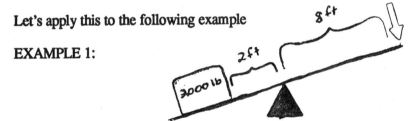

A 2,000-pound (1 ton) weight is placed two feet from the fulcrum of the plank. How much effort would an individual have to exert on the opposite end of the plank, exactly 8 feet from the fulcrum to balance the load?

First, let's identify specifics as they relate to the equation. The 2,000-pound weight represents the *RESISTANCE TO LIFT* and the 2-foot space between the weight and fulcrum is considered the *RESISTANCE DISTANCE*. We are also told that the space between the fulcrum and where a force will be exerted is 8 feet. This is the *EFFORT DISTANCE*. The unknown as represented by X is the amount of force necessary to lift the weight.

If we plug this information into the equation, we are left with a simple proportion to figure:

$$X \times 8 \text{ feet} = 2{,}000 \text{ pounds} \times 2 \text{ feet}$$

To determine X, divide *EFFORT DISTANCE* into the product of the *RESISTANCE* and the *RESISTANCE DISTANCE*.

$$X = \frac{2{,}000 \text{ pounds} \times 2 \text{ feet}}{8 \text{ feet}}$$

$$X = 500 \text{ pounds}$$

Therefore, 500 pounds of force would be required. Given the proper conditions, you can see the dramatic results of leverage as opposed to trying to lift the 2,000-pound weight directly.

Let's look at one more example of leverage and determine the effort required to lift the heavy object.

EXAMPLE 2

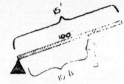

A wooden beam 15 feet long and weighing 100 pounds needs to be moved. If lift is applied 10 feet from the fulcrum, how much effort is required to lift the beam at that point?

Again, identify the specifics as they relate to the equation discussed earlier. The weight of the beam itself, 100 pounds, is the *RESISTANCE TO LIFT*. Since the fulcrum is at the base of the beam, it would be an easy assumption to make that the beam's entire length of 15 feet would serve as the *RESISTANCE DISTANCE*. However, the beam's center of mass must be taken into consideration. Since the beam would be the heaviest to lift at its middle versus either end, we must only figure ½ the beam's length as the true *RESISTANCE DISTANCE*. Therefore, if we place these numbers into the equation, the lift requirement will easily be determined:

$$X \times 10 \text{ feet} = 100 \text{ pounds} \times 7.5 \text{ feet}$$

$$X = \frac{100 \text{ pounds} \times 7.5 \text{ feet}}{10 \text{ feet}}$$

$$X = 75 \text{ pounds of effort required to lift the beam.}$$

Note that the mechanical advantage of Example 1 far exceeds that seen in Example 2. The location of the fulcrum plays a crucial role in leverage effectiveness.

2. PULLEYS

Pulleys are another means of gaining mechanical advantage and making a job easier. Pulleys can be used in a wide variety of ways to pull or lift heavy objects. They can be either in a fixed or movable position. For an illustration, look at the example to the right.

The pulley apparatus shown is depicted lifting a 100-pound weight.

Since the weight is evenly distributed between cables C and D, each cable would be responsible for supporting 50 pounds. Therein lies the mechanical advantage, because only 50 pounds of pulling force is needed at point E to lift the weight. Therefore, the mechanical advantage in quantitative terms is 2:1 (*RESISTANCE* divided by *EFFORT*, i.e., 100 ÷ 50).

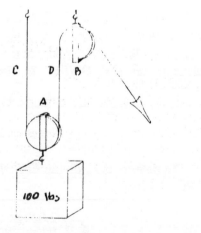

Let's look at another arrangement of pulleys and see if you can determine how much force will be required to lift the weight and what kind of mechanical advantage is gained by the pulleys.

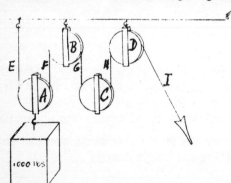

Since the weight is evenly distributed between cables E, F, G, and H, each cable would be responsible for supporting 250 pounds. Thus, it would require only 250 pounds of pulling force to lift the 1,000-pound weight. If you figured that the mechanical advantage was 4 you were correct. (Resistance of 1,000 pounds divided by 250 pounds of effort).

What should be apparent at this point is that what is gained in force, we sacrifice in distance or vice versa. This basic law applies to all forms of simple machinery. Let's look at the sample pulley arrangements below to see this principle at work:

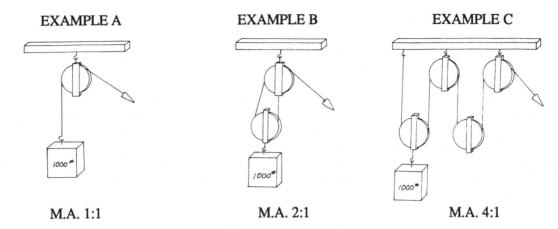

EXAMPLE A — M.A. 1:1
EXAMPLE B — M.A. 2:1
EXAMPLE C — M.A. 4:1

Example A does not have any mechanical advantage besides changing the direction the cable can be pulled. It would require 1,000 pounds of pull to lift the 1,000 pounds of weight. If the cable is pulled 2 feet, the weight will be lifted 2 feet. This is a 1 to 1 proportion.

Example B demonstrates a mechanical advantage of 2:1. Therefore, it would only require 500 pounds of pull to lift the 1,000 pound weight. However, if the cable were to be pulled 2 feet, the 1,000 pounds weight would be raised only 1 foot.

Example C illustrates a pulley system that has a mechanical advantage of 4:1. It requires only 250 pounds of pull to lift the 1,000 pound weight. In this example, if the cable was pulled 2 feet, the 1,000 pound weight would be lifted only half a foot.

The correlation made apparent by these examples is that:

$$\text{EFFORT} = \frac{\text{WEIGHT (Resistance)}}{\text{MECHANICAL ADVANTAGE}}$$

AND

$$\text{LENGTH OF PULL} = \text{HEIGHT (LIFT)} \times \text{MECHANICAL ADVANTAGE}$$

3. INCLINED PLANES

If a truck driver had an extremely heavy barrel (300 lbs.) to load, he would be considered foolish to attempt picking it up directly. Chances are the amount of force required to do that would exceed the driver's physical limitations. Back injury or rupture could result. On the other hand, if a ramp were used to load the barrel into the truck, significantly less effort would be required. Using an inclined plane such as the ramp gives the driver a distinct mechanical advantage.

To quantify this advantage, the length of the inclined plane (effort distance) must be divided by the height of the inclined plane (resistance distance). So if the plank is 15 feet long and the truck bed is 5 feet off the ground, the mechanical advantage would be: 15 feet ÷ 5 feet, or 3:1.

To determine the amount of force required to load the barrel, use the same kind of formula applied for solving leverage problems.

EFFORT x LENGTH OF INCLINED PLANE (EFFORT DISTANCE) = RESISTANCE x HEIGHT (RESISTANCE DISTANCE)

EFFORT x 15 feet = 300 pounds x 5 feet

$$\text{EFFORT} = \frac{300 \text{ pounds} \times 5 \text{ feet}}{15 \text{ feet}} = 100 \text{ lbs}$$

Therefore, 100 pounds of effort is required to load the barrel.

4. WEDGES

In principle, wedges are similar to inclined planes. The only difference is that we use a wedge to push under or between objects that are to be moved instead of moving objects up an incline to gain mechanical advantage. A wood-splitting wedge is a good example. As the wedge is driven further into the end of the big log, more pressure is exerted against the grain, causing the log to eventually split.

5. WHEEL AND AXLE

In general terms, a larger wheel is connected to a rod or axle that performs work, or which turns a smaller wheel that performs work. The force required to turn the larger wheel connected to the smaller wheel or axle bearing resistance is significantly less that required to turn the smaller wheel or axle alone. Therein lies the distinct mechanical advantage that a wheel and axle can provide.

A good example of this would be the crank mechanism or windlass positioned above an open water well. One end of the coiled rope is secured to the axle itself, while the other end is tied to a large pail. As the axle is turned by the attached crank to unwind the coil of rope, the pail is lowered into the well to retrieve water. When the crank is turned in the opposite direction, the pail of water which represents the resistance is hoisted to the surface. Outside of the advantage of directing pull in a true vertical direction (that is, the pail is less prone to drag against the sides of the well), the lift required to hoist the pail of water is significantly less.

To measure the amount of force gained by using the wheel and axle, the circumference of the wheel and axle must be given along with the weight or resistance it is intended to pull. If an axle has a diameter of 4 inches, its circumference would be found by using the following geometrical equation:

CIRCUMFERENCE = π (3.1416) x DIAMETER

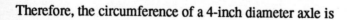

OR

π (3.1416) x RADIUS (½ the diameter) x 2

Therefore, the circumference of a 4-inch diameter axle is

π (3.1416) x 4 inches = 12.566 inches

Once the circumferences of a wheel and axle are known, the following equation can be applied as a proportion to determine force or effort requirements to gain lift:

Force x Circumference of the larger wheel = Resistance x Circumference of the axle or smaller wheel.

For example:

How much force would be required to raise a pail of water weighing 40 pounds if the crank turns a wheel with a circumference of 30 inches and the axle it connects to has a diameter of 6 inches.

Since the circumference of the axle must first be determined, we multiply π (3.1416) with 6 inches to get 18.849 inches.

Now that both circumferences are known, we can easily figure the lift requirement using our equations.

$$X \times 30 \text{ inches} = 40 \text{ pounds} \times 18.849 \text{ inches}$$

$$X = \frac{40 \text{ pounds} \times 18.849 \text{ inches}}{30 \text{ inches}}$$

$$X = 25.13 \text{ pounds of force}$$

The mechanical advantage using this kind of device would be calculated by simply dividing the *RESISTANCE* by the *FORCE* or *EFFORT* required.

$$40 \text{ pounds} \div 25.13 \text{ pounds} = 1.59:1$$

6. SCREWS

Screws can be thought of as being very similar to wedges. They can be used to wedge into a variety of materials such as wood, plastics, and sheet metal, or harnessed to lift heavy objects as demonstrated by the use of a jack screw. Screws are essentially rods with spiral threading. The distance between each two winds of the thread is called the *pitch*. If the screw is turned one complete revolution, the distance that the screw moves or performs work is determined by the pitch. To measure the mechanical advantage of a screw, the circumference of the circle through which the lever which moves the screw turns is divided by the pitch. For example:

If a jackscrew handle is 16 inches long and the screw pitch is ⅙ of an inch, what is the mechanical advantage?

First, we need to determine the potential circumference of the lever (handle) involved. Since the length of the handle is the radius of its turning circle, twice that length is the diameter. This product is multiplied by π (3.1416) to obtain the circumference.

$$16 \text{ inches} \times 2 \times \pi (3.1416) = 100.53$$

Divide this number by the screw pitch to determine the mechanical advantage.

$$100.53 \div \tfrac{1}{6}$$

Note: When dividing by a fraction, multiply both the numerator and the denominator by the inverse of that fraction as shown below:

$$\frac{100.53 \times 6}{\tfrac{1}{6} \times 6} = \frac{603.18}{1} \text{ or } 603:1$$

Now, if we know what kind of mechanical advantage is afforded us by using this jackscrew, it is simple to calculate its potential lift (that is, resistance). Let's assume we exert a 15-pound effort on the handle of the jackscrew. How much weight (that is, resistance) can be lifted by the jack? Since the mechanical advantage has already been determined to be 603:1, we simply set it up in the following proportion:

$$\frac{\text{RESISTANCE } X}{\text{EFFORT OR FORCE (15 POUNDS)}} = 603$$

$$X = 603 \times 15 \text{ pounds}$$

$X = 9,045$ pounds can be lifted by the jackscrew under the conditions given.

II. GEARS AND GEARING

Gears are essentially wheels with teeth; they are normally attached to a shaft. Depending on the origin of the torque (that is, the twisting or torsional force imparted on a driveline) a gear may either be turned by the shaft or may turn the shaft itself. A gear is basically a rotating lever. It can increase or decrease torque by applying the same principle of leverage as discussed earlier or it can change the direction of a force.

There are two general classes of gears, external and interior gears. External gears are the most common and consist of a wheel with exterior teeth. Interior gears, on the other hand, have teeth on the inside to accommodate a smaller gear for torque enhancement.

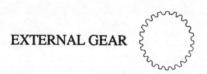

EXTERNAL GEAR   INTERIOR GEAR

Teeth alignment serve to further classify gears. If a gear has straight teeth (that is, teeth set perpendicularly to a gear's facing), it is considered to be a *spur gear*. If the teeth are cut at an angle, this is a *helical gear*. Herringbone gears are two helical gears placed back to back. These more complex designs tend to alleviate vibration and noise.

SPUR   HELICAL   HERRINGBONE

If a drivetrain must change direction, beveled gears are used. The teeth are not situated like those seen on an exterior spur gear. Rather, the teeth are cut into the edging of a gear to give it a beveled appearance.

When we examine gears to see what they do, we can refer to the principles discussed in the section on the wheel and axle. The larger the wheel, the more twist or torque it can apply to a smaller wheel or axle. The same can be said of gears. A larger gear can exert far more torque on the shaft it is connected to than a smaller gear can. However, when we examine torque output between meshing gears, an inverse relationship is seen.

For instance, if the diameter of the drive gear (that is, pitch diameter) is twice the size of a connected smaller gear, the output torque is halved. If the drivegear is three times the size of the second gear, the torque would only be one third as much. On the other hand, if the drive gear is half the size of the second gear, the output torque is doubled. The underlying principle here is that what is sacrificed in torque is gained in speed and vica versa.

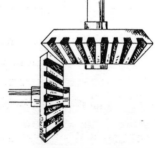

BEVELED GEAR

2 times the torque at ½ the speed

3 times the torque, at ⅓ the speed

½ the torque double the speed

D.G.—*Drive Gear*

Instead of looking at pitch diameter to determine gear ratios, the number of teeth on each gear can be counted and compared. If a drivegear has 8 teeth and the second connecting gear (that is, the driven gear) has

24 teeth, we can determine that the drivegear would have to turn 3 times to turn the driven gear once. Or, in other terms, the gear ratio (i.e., mechanical advantage) is 3:1 and the torque generated would be 3 times as much. As a basic guideline, the more teeth (hence larger pitch) a given gear has, the slower it will run resulting in a proportionate increase in torque.

When looking at several gears in a driveline, it can be a little confusing to try and determine the direction in which any one gear will turn. It is understood that when one gear meshes with another, the direction of the driven gear is always opposite that of the drive gear.

When several gears are shown in a driveline, the principle remains the same. Remembering which gear went in what direction may get a little confusing toward the end of the drivetrain; here is a simple solution to that problem: Number each gear as it falls in succession, such as shown in the example below:

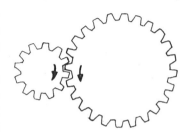

The odd-numbered gears will always turn in one direction while the even-numbered gears turn in the opposite direction. The exception to this rule is an internal gear (Gear #8 in the example shown). It always turns in the same direction as the gear surrounding it.

Now, if we desire to quantify how much of an effect the driver gear will have on the last gear of a drivetrain, the following procedure can help:

Let G_1 represent the first gear (driver gear)

Let G_L represent the last gear in the drivetrain

Let P_D represent the product of the number of teeth involved to cause drive

Let P_X represent the product of the number of teeth being driven

Let's apply this format to the drivetrain illustrated to the right:

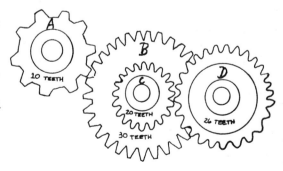

(Note: Gears B and C are situated on the same shaft. Therefore, the direction is the same for both.)

For every one revolution of gear A, how much of a revolution will gear D make?

G_L is our unknown

G_1 demonstrates one revolution (1)

P_D the driver gears are A and C. They have 10 and 20 teeth, respectively. Therefore, the product would be 10 x 20 = 200.

P_X the gears being driven are B and D. They have 30 and 26 teeth respectively. Therefore, the product would be 30 x 26 = 780.

$$G_L = G_1 \times \frac{P_D}{P_X}$$

Using this equation, we can plug in our known factors and determine G_L.

$$G_L = (1 \text{ revolution}) \times \frac{10 \times 20}{30 \times 26}$$

$$G_L = 1 \times \frac{200}{780}$$

$$G_L \quad \frac{1}{3.90}$$

So, in other words, this particular gear reduction causes the last gear to turn almost ¼ of a revolution for every 1 revolution made by the first driver gear. This demonstrates a mechanical advantage of 4:1 or 4 times as much torque due to the configuration of gears involved.

This is a very simplified overview of gears and their applications. There is a host of technical information that can be discussed relating to this topic. However, the most important concept that should be learned here, is that there is no such thing as perpetual motion. What we achieve in force, we lose in distance; whatever torque is gained, we lose in speed. This is assuming the source of power (e.g., an engine) is constant.

III. BELT DRIVES

Belt drives are very comparable to gears, with the exception of the manner in which they are interconnected to perform work. Instead of various sized gears directly intermeshing, wheels are connected by means of a belt. This too, depending on the configuration of wheels, can increase or decrease speed or torque. If the belt is twisted in the manner shown below, it can directly change the direction of the opposing wheel.

In general, the mechanical principles discussed for gears and their applications are the same for belt drives.

IV. HYDRAULICS AND FLUID DYNAMICS

Fluid dynamics is an extremely complicated field in itself. However, there are a few basic properties that apply to all fluids and their mechanical uses. To begin with, fluid pressure always acts on a given surface in a perpendicular manner. A rod in the cylinder of a hydraulic apparatus would be a good example. As pressure is increased via the inlet port, there is a corresponding perpendicular force exerted on the piston to perform work.

Second, if pressure is exerted on a fluid trapped in a confined space, the pressure is distributed in all directions without consequent loss of power used to create the pressure.

Let's look at the hydraulic ram, again. If the piston encounters an opposing force equal to that which is coming through the inlet port, the trapped fluid will exert an equal pressure at any point within its container.

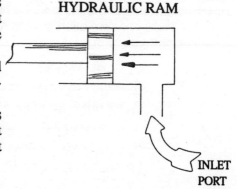

HYDRAULIC RAM

INLET PORT

MECHANICAL APTITUDE

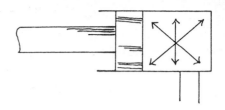

A third property is that fluid pressure is the same at a given depth in all directions as well as being proportional to that depth (that is, in an open container). For example, if a scuba diver descends to a depth of 33 feet, regardless of where that diver swims on that plane, he or she will have to tolerate an ambient pressure of 29.4 pounds per square inch. (At sea level, it is only 14.7 pounds per square inch). At 66 feet, it would be 44.1 pounds per square inch. At 99 feet, it would be 58.8 pounds per square inch, and at 132 feet, it would be 73.5 pounds per square inch.

Varying fluid densities play a role, too. Salt water is actually heavier than fresh water (64 pounds per cubic foot versus 62.4 pounds per cubic foot) because of the salts in suspension. Therefore, a diver at a depth of 33 feet in fresh water is subject to less pressure than a diver in 33 feet of salt water. The proportional changes of pressure as dictated by depth still occur regardless of the fluid density involved.

Finally, fluid pressure is unaffected by the dimension of open containment. For example, the ocean floor may be six miles deep in a given area. The pressure for that depth is the same regardless of the expanse of shallower water on its periphery.

These few examples demonstrate the basic guidelines that affect fluid dynamics. It is important that they be understood because they play a key role in hydraulic mechanics. Some of the sample questions that follow will test your familiarity with fluid properties. You will probably encounter similar questions on the actual exam.

SAMPLE QUESTIONS ON MECHANICAL PRINCIPLES

1. If someone used a 10-foot-long lever to move a 2½ ton rock, which location of the fulcrum would provide the greatest lift?
    A. 2 feet from the rock
    B. 4 feet from the rock
    C. 6 feet from the rock
    D. The end of the lever held by the user

2. If a lever were used to move a crate filled with heavy machinery parts under the following conditions, how much effort would be required to lift it? The crate weighs 3,000 pounds and measures 3½ feet by 4 feet. The fulcrum was placed 2 feet from the crate. The lever itself is 10 feet long. Effort will be applied at the very end of the lever opposite the load.
    A. 675 pounds   B. 700 pounds   C. 725 pounds   D. 750 pounds

3. Look at question 2 again. Under the described conditions, what mechanical advantage is gained by using the lever as opposed to lifting the crate directly?
    A. 1:4   B. 3:2   C. 4:1   D. 2:3

PRY RESISTANCE IS 500 LBS

4. If a wrecking bar is used to pry apart laminated framework as shown in the above diagram, how much effort would be required to do the job?
    A. 50 pounds   B. 60 pounds   C. 65 pounds   D. 70 pounds

5. Three people are struggling to raise a 30-foot flagpole that measures 3 inches in diameter. The total weight of the flagpole is 95 pounds. If one person served as the fulcrum by anchoring the end of the flagpole to the ground, and the other two people attempted to lift the flagpole at a point 12 feet from the fulcrum, how much resistance would they encounter?
   A. 95 pounds   B. 160.35 pounds   C. 118.75 pounds   D. 250 pounds

6. Refer again to question 5. What mechanical advantage is afforded by lifting the flagpole as described?
   A. 1:1
   B. 2:1
   C. Can't be determined with the information provided
   D. There is no advantage gained by using leverage in this manner

7. Which pulley can offer the most amount of lift?

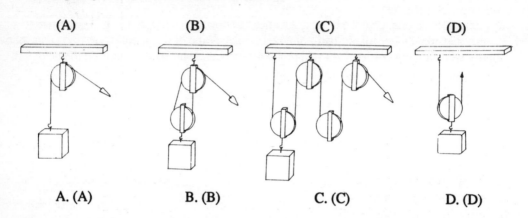

   A. (A)   B. (B)   C. (C)   D. (D)

8. Of the pulley arrangements shown in question 7, which one does not provide any mechanical advantage?
   A. (A)   B. (B)   C. (C)   D. (D)

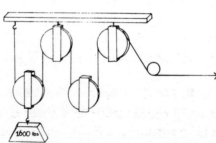

9. Examine the above diagram. How much effort would be required, using this set of pulleys, to lift the 1,600-pound weight?
   A. 300 pounds   B. 400 pounds   C. 500 pounds   D. 550 pounds

10. Refer to the pulley diagram in question 9. If the cable were pulled 10 feet, how high would the 1,600-pound weight rise?
    A. 2 feet   B. 2.5 feet   C. 3.0 feet   D. 3.5 feet

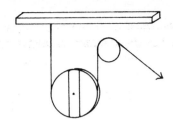

11. Refer to the diagram above. If 600 pounds of pulling force were exerted on the cable, what would be the maximum load this pulley could lift?
    A. 600 pounds   B. 1,000 pounds   C. 1,200 pounds   D. 2,400 pounds

12. How much pulling force will be required to lift the 1,000 pounds of resistance in the diagram to the right?

    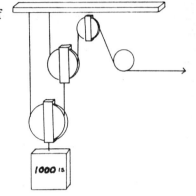

    A. 1,000 pounds   B. 500 pounds

    C. 350 pounds   D. 250 pounds

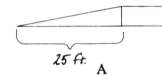 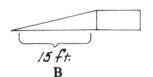 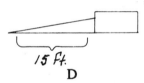

      A   B   C   D

13. Which of the inclined plane diagrams above demonstrates the best mechanical advantage in lifting a heavy object to the height of the platform?
    A. (A)   B. (B)   C. (C)   D. (D)

14. What is the mechanical advantage of diagram B shown in question 13, if the height of the platform is 10 feet?
    A. 2.5 : 1   B. 2.0 : 1   C. 1.75 : 1   D. 1.5 : 1

15. If someone needed to load a 425-pound fuel barrel into the back of a truck with a bed height of 4 feet, how much effort would be required if a 10-foot plank were used as an inclined plane?
    A. 150 pounds   B. 170 pounds   C. 180 pounds   D. 185 pounds

16. What would be the height that a 500 pound barrel could be lifted if a 20-foot inclined plank having a mechanical advantage of 2:1 was used and only 150 pounds of force is available?
    A. 4 feet   B. 5 feet   C. 6 feet   D. 7 feet

17. What is the mechanical advantage of the inclined plane shown in the diagram to the right?

    A. 2.6 : 1   B. 2.4 : 1   C. 2.2 : 1   D. 2.0 : 1

COMPLETE FIRE-FIGHTERS EXAM PREPARATION BOOK

18. Examine the diagram to the right. If a crank handle is connected to a 5-inch-diameter axle and has a turning circumference of 20 inches, how much force would be required to lift 300 pounds?

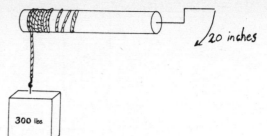

   A. 245.75 pounds   B. 240.15 pounds

   C. 237.80 pounds   D. 235.65 pounds

19. What mechanical advantage is gained by using the crank shown in question 18?
   A. 2.21 : 1   B. 1.27 : 1   C. 1.15 : 1   D. 1 : 1

20. How much weight could be lifted if 150 pounds of force were exerted on the crank connected to an axle with a known circumference of 8.5 inches, as shown in the diagram to the right?

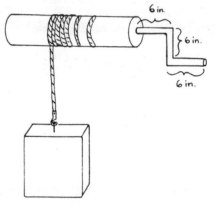

   A. 615.27 pounds   B. 631.65 pounds

   C. 637.65 pounds   D. 665.29 pounds

21. What is the mechanical advantage of the windlass described in question 20?
   A. 3.71 : 1   B. 4.22 : 1   C. 4.44 : 1   D. 4.71 : 1

22. Let's assume a ⅜-inch hole is drilled into a block of steel and later tapped to create a ⅛-inch pitch. If a ⅜-inch bolt with a ⅛-inch pitch is screwed into the hole with a box-end wrench 6 inches long, what is the potential force of the bolt as measured at its threaded end if we apply 10 pounds of force?
   A. 2,576 pounds of force
   B. 3,016 pounds of force
   C. 3,758 pounds of force
   D. 3,982 pounds of force

23. If a jackscrew with a pitch of ½ inch possesses a handle (lever) 20 inches long to which 10 pounds of force is applied, how much potential weight could be lifted?
   A. 8,796 pounds   B. 8,875 pounds   C. 9,257 pounds   D. 9,762 pounds

24. What mechanical advantage is gained by using the jackscrew in question 23?
   A. 879.62 : 1   B. 887.51 : 1   C. 925.70 : 1   D. 976.29 : 1

25. How much force or effort would be required to lift 3 tons using a jackscrew with a ⅙-inch pitch and a lever/handle 18 inches long?
   A. 6.0 pounds   B. 7.62 pounds   C. 7.75 pounds   D. 8.84 pounds

26. Examine the diagram below. Which statement is correct, assuming that the drive gear is turning counterclockwise?

A. All the gears are turning counterclockwise.
B. Gears 1, 3, and 6 all turn counterclockwise.
C. Gears 2, 4, and 6 all turn clockwise.
D. Since gear 4 is an internal gear, it rotates in the same direction as gear 3.

27. What kind of gears are used in changing the direction of a drivetrain?
A. External spur B. Beveled gears C. Helical gears D. Herringbone gears

28. If a drivegear has 30 teeth and a connecting gear (that is, driven gear) has 10 teeth, which of the following will occur?
A. The output torque is tripled while speed is cut by ⅓.
B. The output torque and speed are tripled.
C. The output torque is only ⅓ while speed is tripled.
D. The output torque is increased only 25%.

29. Refer to the diagram to the right. Assuming gears A and B have 10 teeth each and gear C has 20 teeth, what would happen if gear B were the drivegear and turned ½ a revolution counterclockwise?

A. Gears A and C would each turn ½ a revolution counterclockwise.
B. Gear A would turn ½ revolution counterclockwise and gear C would turn clockwise.
C. Gears A and C would respectively turn ¼ revolution and ½ revolution clockwise.
D. Gears A and C would respectively turn ½ and ¼ revolution clockwise.

30. Refer to the diagram to the right.
A has 10 teeth, B has 40 teeth, C has 10 teeth, D has 20 teeth. If Gear A turned one full revolution, how much would Gear D turn?
A. ⅕th of a revolution
B. ⅛th of a revolution
C. ⅑th of a revolution
D. ⅒th of a revolution

NOTE: Gears B and C are keyed to the same shaft.

31. Refer to the diagram to the right.

    A has 10 teeth
    B has 40 teeth
    C has 20 teeth
    D has 30 teeth

    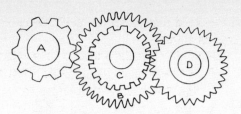

    NOTE: Gears B and C are keyed to the same shaft.

    If Gear A turned 6 revolutions, how much would Gear D turn?
    A. 1 revolution   B. 1.5 revolutions   C. 1.75 revolutions   D. 2 revolutions

32. Assuming that a V-8 350 engine could yield 250 horsepower creating a maximum torque of 175 foot-pounds, and the gear configuration demonstrated in question 31 was its drivetrain, what would be the maximum torque output of gear D?
    A. 1,000 foot-pounds
    B. 1,050 foot-pounds
    C. 1,100 foot-pounds
    D. 1,150 foot-pounds

33. Belt drives can do which of the following?
    A. Decrease torque
    B. Decrease speed
    C. Change drive direction
    D. All of the above

QUESTIONS 34 THROUGH 37 RELATE TO THE DIAGRAM SHOWN BELOW

Wheel dimensions are as follows:

   A—10 inches in diameter
   B—10 inches in diameter
   C—5 inches in diameter
   D—10 inches in diameter
   E—15 inches in diameter

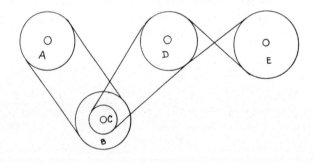

34. If wheel A turns at a rate of 2,000 rpm (revolutions per minute), how fast would wheel C turn?
    A. 1,000 rpm   B. 2,000 rpm   C. 2,500 rpm   D. None of the above

35. If wheel A turns at a rate of 5,000 rpm, how fast will wheel D turn?
    A. 5,000 rpm   B. 3,000 rpm   C. 2,500 rpm   D. 1,500 rpm

36. If wheel A is turned clockwise, which of the following statements is true?
    A. Wheel E would turn counterclockwise.
    B. Wheels D and E would turn counterclockwise.
    C. Wheels A and D would turn clockwise and wheels C and E counterclockwise.
    D. Wheels B, C, D, and E would turn counterclockwise.

37. If wheel D is the driver wheel and it makes 4 counterclockwise revolutions, which of the following statements is true?
    A. Wheel E will make 4 clockwise revolutions.
    B. Wheel C will make 4 clockwise revolutions.
    C. Wheel C will make 4 counterclockwise revolutions.
    D. Wheel A will make 8 counterclockwise revolutions.

38. If the piston exerted a pressure of 200 pounds per square inch (PSI) on the water trapped in the pipe (i.e., a closed system), which of the following statements is correct?

    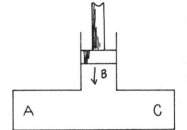

    A. The pressure would be the greatest at point B.
    B. The pressure is the same throughout the closed system.
    C. The pressure measured at points A and C is twice that at point B.
    D. The pressure is diminished at both points A and C.

39. Refer to the diagram below.

    A   B   C

    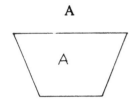 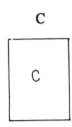

    Assuming the base of each container is the same and that each container is full of salt water, which of the three vessels shown would have the greatest pressure at its base?
    A. (A)   B. (B)   C. (C)   D. All vessels would have the same pressure at their base because the same kind of fluid was used (that is, fluid density is the same).

40. Which of the following diagrams demonstrate the quickest way to siphon the fluid in the container, assuming all things are equal except the length of the siphon tube itself?

    A   B   C

      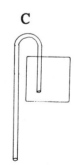

    A. (A)   B. (B)   C. (C)   D. Cannot be determined with the information given.

ANSWER SHEET FOR SAMPLE MECHANICAL PRINCIPLES QUESTIONS

1. Ⓐ Ⓑ Ⓒ Ⓓ
2. Ⓐ Ⓑ Ⓒ Ⓓ
3. Ⓐ Ⓑ Ⓒ Ⓓ
4. Ⓐ Ⓑ Ⓒ Ⓓ
5. Ⓐ Ⓑ Ⓒ Ⓓ
6. Ⓐ Ⓑ Ⓒ Ⓓ
7. Ⓐ Ⓑ Ⓒ Ⓓ
8. Ⓐ Ⓑ Ⓒ Ⓓ
9. Ⓐ Ⓑ Ⓒ Ⓓ
10. Ⓐ Ⓑ Ⓒ Ⓓ
11. Ⓐ Ⓑ Ⓒ Ⓓ
12. Ⓐ Ⓑ Ⓒ Ⓓ
13. Ⓐ Ⓑ Ⓒ Ⓓ
14. Ⓐ Ⓑ Ⓒ Ⓓ
15. Ⓐ Ⓑ Ⓒ Ⓓ
16. Ⓐ Ⓑ Ⓒ Ⓓ
17. Ⓐ Ⓑ Ⓒ Ⓓ
18. Ⓐ Ⓑ Ⓒ Ⓓ
19. Ⓐ Ⓑ Ⓒ Ⓓ
20. Ⓐ Ⓑ Ⓒ Ⓓ
21. Ⓐ Ⓑ Ⓒ Ⓓ
22. Ⓐ Ⓑ Ⓒ Ⓓ
23. Ⓐ Ⓑ Ⓒ Ⓓ
24. Ⓐ Ⓑ Ⓒ Ⓓ
25. Ⓐ Ⓑ Ⓒ Ⓓ
26. Ⓐ Ⓑ Ⓒ Ⓓ
27. Ⓐ Ⓑ Ⓒ Ⓓ
28. Ⓐ Ⓑ Ⓒ Ⓓ
29. Ⓐ Ⓑ Ⓒ Ⓓ
30. Ⓐ Ⓑ Ⓒ Ⓓ
31. Ⓐ Ⓑ Ⓒ Ⓓ
32. Ⓐ Ⓑ Ⓒ Ⓓ
33. Ⓐ Ⓑ Ⓒ Ⓓ
34. Ⓐ Ⓑ Ⓒ Ⓓ
35. Ⓐ Ⓑ Ⓒ Ⓓ
36. Ⓐ Ⓑ Ⓒ Ⓓ
37. Ⓐ Ⓑ Ⓒ Ⓓ
38. Ⓐ Ⓑ Ⓒ Ⓓ
39. Ⓐ Ⓑ Ⓒ Ⓓ
40. Ⓐ Ⓑ Ⓒ Ⓓ

ANSWERS TO SAMPLE MECHANICAL PRINCIPLE QUESTIONS

1. **A.** The closer a fulcrum is moved to a heavy object (resistance), the easier it is to apply leverage to move the object.

2. **D.**
$$\text{Effort} = \frac{3{,}000 \text{ pounds} \times 2 \text{ feet}}{8 \text{ feet } \{10 \text{ feet} - 2 \text{ feet}\}} = 750 \text{ pounds needed}$$

3. **C.** Mechanical advantage is arrived at by dividing resistance by the amount of force required to lift it.

$$\frac{3000 \text{ pounds}}{750 \text{ pounds}} = 4, \text{ or a 4:1 mechanical advantage}$$

4. **B.**
$$\text{Effort} = \frac{500 \text{ pounds} \times 3 \text{ inches}}{25 \text{ inches}} = 60 \text{ pounds of force required to pry the framework apart.}$$

5. **C.** Since the fulcrum point is at the base of the pole, the entire length of the flagpole is considered the resistance distance. Therefore,

$$\text{Effort} = \frac{95 \text{ pounds} \times 15 \text{ feet}}{12 \text{ feet}}$$

Effort = 118.75 pounds of force is required at that point to lift the flagpole

6. **D.** Since mechanical advantage is determined by dividing resistance by the amount of force required to lift

$$\frac{(95 \text{ pounds})}{(118.75 \text{ pounds})}$$

we arrive at a mechanical advantage of less than 1 (0.80) In other words, leverage is actually working against the two individuals trying to raise the flagpole. In fact, the closer they come to the fulcrum to attempt lift, the greater the resistance to their effort will be.

7. **C.** The pulley apparatus shown in choice C has a mechanical advantage of 4 : 1. Therefore, it can lift more weight as compared to the others; they have smaller mechanical advantages.

8. **A.** The pulley shown in choice A affords no mechanical advantage. It only changes the direction of pull. If you were to lift 500 pounds of weight, it would require 500 pounds of pull or effort. Choices B and D both offer a mechanical advantage of 2. If 500 pounds were lifted by either of these two pulley arrangements, the effort required would be only 250 pounds.

9. **B.** Since this example shows a pulley that has a mechanical advantage of 4:1, we simply divide the weight (that is, resistance) by the mechanical advantage to obtain the amount of effort required to lift it.

$$\frac{1600}{4} = 400 \text{ pounds of effort required}$$

10. **B.** Length of pull = Height (Lift) x Mechanical Advantage

    OR

    $$\frac{\text{Length of pull}}{\text{Mechanical Advantage}} = \text{Height (Lift)}$$

    $$\frac{10 \text{ feet}}{4} = 2.5 \text{ feet of lift}$$

11. **C.** The pulley has a 2:1 mechanical advantage. Therefore,

    600 pounds of force x 2 = 1,200 pounds of potential lift.

12. **D.** The mechanical advantage demonstrated by this kind of pulley configuration is 4:1. The smaller wheel closest to the source of pull serves only to change the direction of that effort. This device is referred to as a snatch block. It does not contribute to mechanical advantage.

    $$\text{Effort} = \frac{\text{Resistance}}{\text{Mechanical Advantage}} = \frac{1000 \text{ pounds}}{4} = 250 \text{ pounds of effort}$$

13. **A.** The mechanical advantage is determined by the length of the plane divided by the height it is elevated. Since choice A has the longest plane and the height is the same for all selections, plane A would have the greater mechanical advantage. Both choices C and D still require direct lift.

14. **D.**

    $$\text{Mechanical Advantage} = \frac{\text{Length of inclined plane}}{\text{Height of inclined plane}}$$

    $$\frac{15 \text{ feet}}{10 \text{ feet}} = 1.5 : 1$$

15. **B.**

    Effort x Length of inclined plane = Resistance x Height

    Effort x 10 feet = 425 pounds x 4 feet

    $$\text{Effort} = \frac{425 \text{ pounds} \times 4 \text{ feet}}{10 \text{ feet}} = 170 \text{ pounds of force needed}$$

16. **C.**

    150 lbs. of effort x 20 feet = 500 lbs. of resistance x height

$$\frac{150 \text{ lbs.} \times 20 \text{ feet}}{500 \text{ lbs}} = 6 \text{ feet}$$

Under the conditions described, the 500-pound barrel can be lifted 6 feet.

17. **A.** Since we must first determine the length of the plane, we have to use the Pythagorean theorem applied to geometric right triangles. That is, if we square the base and height and add them together, the square root of the resulting number will give us the length of the third side (that is, the inclined plane).

$$5^2 = 25$$
$$144 + 25 = 169; \quad \sqrt{169} = 13$$
$$12^2 = 144$$

$$\text{Mechanical advantage} = \frac{\text{Length of plane}}{\text{Height}} = \frac{13 \text{ height}}{5} = 2.6 : 1$$

18. **D.** First we need to determine the circumference of the axle.

$$\pi (3.1416) \times 5 \text{ inches} = 15.7075$$

We round to 15.71 inches. With both circumferences known, we use the following equation to determine force requirements.

Force x Large-wheel circumference = Resistance x Axle circumference

Force x 20 inches = 300 lbs. x 15.71 inches

$$\text{Force} = \frac{300 \text{ lbs.} \times 15.71 \text{ inches}}{20 \text{ inches}} = 235.65 \text{ lbs. of force}$$

19. **B.** Divide the resistance by the force required to lift it.

$$\frac{300 \text{ lbs.}}{235.65 \text{ lbs.}} = 1.27 : 1$$

20. **D.** We need to first determine the crank handle's turning circumference. The length of handle that protrudes from the axle's axis is not important. What is important is the measurement of that portion of the handle that constitutes the radius of circular movement. Disregarding the other measurements then, we are left with the length of handle that will determine the potential turning circumference. When a radius is given, we need to apply the following equation to determine circumference:

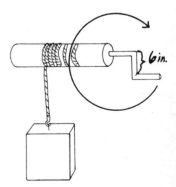

π (3.1416) x 2 (radius is only ½ the diameter) x 6 inches

Circumference of the handle is then equal to 37.70 inches.

Using the same equation applied in question 18.

150 lbs. of force x 37.7 inches = Resistance (wt) x 8.5 inches

$$\frac{150 \text{ lbs. of force} \times 37.7 \text{ inches}}{8.5 \text{ inches}} = \text{Resistance} = 665.29 \text{ lbs.}$$

Therefore, the amount of weight that can be lifted by this crank using 150 lb. of force is 665.29 lbs.

21. *C.* Divide the resistance by the amount of force used to determine mechanical advantage.

$$\frac{665.29 \text{ lbs. of weight}}{150 \text{ lbs. of force}} = 4.44 : 1$$

22. *B.* The circumference of the lever—in this case, the box-end wrench—needs to be figured first. Since six inches represents only the radius of a circle, the following format should be used.

6 inches x 2 x π (3.1416) = 37.7 inch circumference

Now, divide this number by the screw pitch to determine the mechanical advantage:

$$\frac{37.7 \text{ inches}}{\text{⅛ inch pitch}} \text{ OR } 37.7 \text{ inches} \times 8 \text{ inches} = 301.60 : 1$$

$$\frac{\text{Resistance } (X)}{\text{Force of 10 lbs. applied}} = 301.60$$

OR

301.60 x 10 lbs. = Resistance = 3,016 lbs. of force

23. *A.* This is basically the same kind of problem as seen in question 22.

20 inches x 2 x π (3.1416) = 125.66 inch circumference

$$\frac{125.66 \text{ inch}}{\text{⅐ inch pitch}} \text{ OR } 125.66 \times 7 \text{ inches} = 879.62 : 1$$

$$\frac{\text{Weight}}{\text{Force of 10 lbs. applied}} = 879.62$$

OR

879.62 x 10 lbs. = 8,796.2 lbs.

8,796.2 lbs. can be lifted by this jackscrew under the conditions given. The mechanical advantage is 879.62 : 1.

24. A. Refer to the explanation given for question 23.

25. D. Tons should be converted to pounds; 3 tons = 6000 lbs. The lever or handle's potential circumference = 18 x 2 x π = 113.09 inches

$$\text{Mechanical advantage} = \frac{113.09}{1/6} \quad \text{OR} \quad 113.09 \times 6 = 678.54$$

$$\text{Weight given} = \frac{6,00 \text{ lbs.}}{(X) \text{ lbs. of effort needed for lift}} = 678.5$$

Therefore, 6,000 lbs. of resistance = 678.54 x lbs. effort.

$$\frac{6000 \text{ lbs.}}{678.54} = 8.84 \text{ lbs. of effort}$$

26. C. Only choice C is correct. Always remember that if external gears such as those seen in a drivetrain are numbered in succession, the odd-numbered gears will all spin in one direction while the even-numbered gears will spin in the opposite direction. Gear 4 is not an internal gear.

27. B. Beveled gears are used for directing power around a corner or angle.

28. C. Since the drivegear has three times as many teeth as the driven gear, there is a ⅔ reduction of torque with a corresponding tripling in speed. None of the other alternative is correct.

29. D. If gear B turns counterclockwise, we know that connecting gears must turn in the opposite direction. Gears A and B have a 1:1 ratio. If gear B turns ½ a revolution, so will gear A. Gear C, on the other hand, has twice the number of teeth as gear B, so it will rotate only half as much. Since gear B made only ½ a revolution, gear C will exhibit a ¼ revolution.

30. B. Gears A and C are the drivegears shown in the diagram. Note that gears B and C are on the same shaft. Applying the formula:

$$G_L = G_1 \times \frac{P_D}{P}$$

G_L is our unknown, the speed of the last gear.

G_1 is gear A, or the first gear shown in the drivetrain. Since it is given in the problem that it makes only 1 revolution, its value is 1.

P_D represents the product of the teeth of gears A and C (i.e., 10 x 10).
P represents the product of the teeth of gears B and D (i.e., 40 x 20).

Therefore:

$$G_L = 1 \times \frac{10 \times 10}{40 \times 20}$$

$$G_L = \frac{100}{800}, \text{ or } \tfrac{1}{8} \text{ of a revolution}$$

31. **A.** Gears A and C are the drivegears; B and D are the driven gears.

    $$G_L = \frac{6 \times 10 \times 20}{40 \times 30}$$

    $$G_L = \frac{6 \times 200}{1200}$$

    $G_L = 6 \times \tfrac{1}{6}$ therefore; $G_L = 1$

32. **B.** Since we know the gear configuration in question 31 is a 1:6 reduction, its mechanical advantage is 6. Therefore,

    6×175 foot pounds $= 1050$ foot pounds

    1,050 foot pounds is the maximum torque output possible under the conditions given.

33. **D.** Depending on how belt drives are set up, they can increase or decrease speed and torque as well as change drive direction.

34. **B.** Wheels B and C are connected to the same shaft. Since B is the driver wheel of the two and has a 1:1 wheel ratio with wheel A, wheels A, B, and C would all spin at the rate of 2,000 rpm.

35. **C.** It was established in question 34 that wheel A turns as fast as wheel C. Since wheel C is half the size of wheel D, the turning rate is reduced 50%.

    $5,000 \div 2 = 2,500$ rpm.

36. **A.** Only choice A is true. When a belt is twisted between two wheels as shown between D and E, the purpose is to reverse direction.

37. **D.** Only choice D is correct. It's true that wheel E will make a clockwise revolution, but it does not have a 1:1 wheel ratio in relation to D.

38. **B.** Pressure exerted on a fluid trapped in a confined space is distributed in all directions without consequent loss of force.

39. **C.** Fluid pressure is directly related to the depth and density of a given fluid. Since the density is the same because salt water is the only fluid involved, we can simply compare depth. Container C has the greatest depth; therefore it will have the greatest pressure at its base.

40. **C.** The basic procedure for determining fluid pressure is to multiply the height (depth) of a fluid and its density. We can not quantify fluid pressure from the information given. However, choice C demonstrates the longest siphon tube of the three choices. Since it has the greatest height/depth, it can siphon fluid at a greater rate than the other two vessels.

DIRECTIONAL ORIENTATION

A superb memory is a real benefit to a firefighter. However, when it is coupled with good directional orientation, the firefighter is an invaluable asset to his or her department. If a firefighter is familiar with the directional layout of a city or district, emergencies can be responded to in a safer and more efficient manner. On the other hand, a misdirected approach to an emergency can waste valuable time and perhaps further endanger the lives of those involved.

Directional orientation may have implications for a firefighter's own safety, too. For instance, a firefighter may be familiar with a given floor plan, but if conditions, as for example smoke, obscure recognizable landmarks, it is imperative that he or she can orient toward known exits. For these reasons, fire departments are interested in determining a test applicant's directional abilities.

Most of the directional orientation questions on past exams have focused on one of three aspects. The first variety of questions may ask what would be the most efficient way of getting from one point on a street map to another point. If there is a vehicle involved, it is assumed that no traffic violations may be committed en route. For example, no U-turns are permitted, and it is prohibited to travel in the wrong direction on a one-way street.

The second type of questions that may be encountered examine how well you can follow explicit directions. A specific route is outlined, and you will need to determine what or where the final destination is by following those directions. This may seem easy, but its surprising how far off in the wrong direction you can go if there is the slightest misinterpretation of any direction given.

The third type of test questions ask the directional location of an object in relation to a person or specific landmark. This could simply come in the form of someone asking directions on how to get to a particular destination, or it may concern the relative location of two different buildings. These types of questions are probably the easiest to solve because you can figure out any direction by looking at the legend of the illustration given (e.g., see compass at right). Even if the legend specifies only one direction as it applies to the map, you can easily extrapolate any other direction as needed (see compass at right).

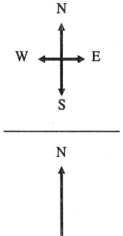

On the exam, a city grid or some form of floor plan will be given from which questions will be asked. Be sure to follow the directions word for word in each question. Outside of not breaking any traffic laws, don't assume anything. It is also important to approach these questions from the perspective of the person asking it. A northern heading from the questioner's standpoint may very well be an altogether different heading for you, depending on the way you look at the map. A right turn for the questioner could mean a left turn for you.

To circumvent this disorientation, simply rotate the diagram to view it from the direction the questioner is looking from. This will eliminate any confusion when directions are discussed. As a final tip, you can more clearly determine correct answers by lightly sketching, on the diagram provided, all proposed or alternative routes in the question. This will allow for better contrast and easier discernment of the correct answer. If there are several questions relating to the same diagram, it is important to erase any previous sketching. Otherwise, you will end up with a confusing array of lines, which can potentially lead to the selection of a wrong answer.

DIAGRAM A

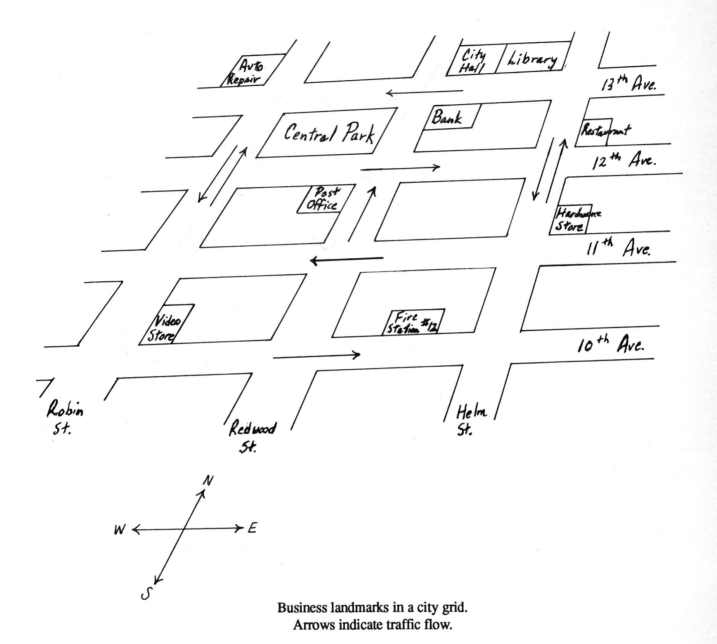

Business landmarks in a city grid.
Arrows indicate traffic flow.

DIRECTIONAL ORIENTATION

COMPLETE FIRE-FIGHTERS EXAM PREPARATION BOOK

SAMPLE QUESTIONS ON DIRECTIONAL ORIENTATION

Answer questions 1 through 6 on the basis of Diagram A

1. If a fire were reported at the bank, what would be the most direct approach for a ladder truck dispatched from the 10th Avenue fire station?
   A. East on 10th Avenue, north on Helm Street, and west on 13th Avenue.
   B. East on 10th Avenue, north on Helm Street, and west on 12th Avenue.
   C. West on 10th Avenue, north on Redwood to 13th Avenue.
   D. East on 10th Avenue, north on Helm Street, west on 11th Avenue, north on Redwood to 13th Avenue.

2. Assuming that each city block has a fire hydrant located in the southeast corner, which fire hydrant would be closest to the bank fire?
   A. The fire hydrant located at the corner of 12th Avenue and Helm Street.
   B. The fire hydrant located at the corner of 12th Avenue and Redwood Street.
   C. The fire hydrant located at the corner of 13th Avenue and Redwood Street.
   D. The fire hydrant located at the corner of 13th Avenue and Helm Street.

3. What building would be in jeopardy from the bank fire if a 30-mph wind were coming from the southwest?
   A. Restaurant   B. Library   C. City Hall   D. Post Office

4. Central Park is bounded by 12th and 13th Avenues and the streets Redwood and Robin. A woman and her pet dog were strolling along the sidewalk at the northeast corner of the park when the dog slipped its leash. If the dog ran southwest across the park and then south at that street intersection for two blocks before heading east for another two blocks, what location would the dog end up being the closest to?
   A. Hardware store   B. Video store   C. Post Office   D. Fire station

5. If the owner of the video store needed to renew a business permit, what would be the most direct route from his place of business to City Hall?
   A. North on Robin and east on 13th Avenue.
   B. East on 10th Avenue, north on Helm, and west on 13th Avenue.
   C. West on 10th Avenue and south on Redwood.
   D. East on 10th Avenue and north on Redwood.

6. Where is the auto repair shop in relation to the fire department?
   A. Southeast   B. Southwest   C. Northeast   D. Northwest

DIRECTIONAL ORIENTATION

DIAGRAM B

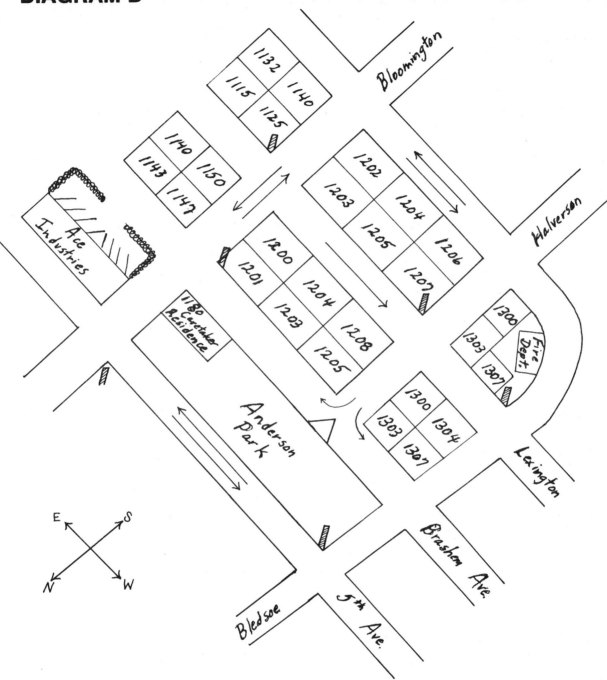

The numbers seen in the diagram indicate residential addresses

▨ represents fire hydrants

∽∽∽ represents fencing

Note: Bledsoe is a heavily traveled thoroughfare for working commuters.

89

COMPLETE FIRE-FIGHTERS EXAM PREPARATION BOOK

Answer questions 7 through 12 on the basis of Diagram B

7. If it were reported that a jogger collapsed one-half block west of the caretaker's residence at Anderson Park, what route should firefighter paramedics take to get to the victim?
   A. North on Bledsoe, then west on Brashem Avenue.
   B. East on Bledsoe, north on Bloomington, and then west on Brashem Avenue.
   C. East on Bledsoe, north on Halverson, and then east on Brashem Avenue.
   D. North on Bledsoe and then east on 5th Avenue.

8. If a strong wind were coming from the northeast and the residence at 1203 Brashem Ave. were on fire, what other residence would be in immediate danger of catching fire?
   A. 1208 Lexington
   B. 1205 Brashem Avenue
   C. 1200 Lexington
   D. 1208 Halverson

9. If the residence at 1201 Brashem Avenue were on fire at approximately 4:45 p.m., what would be the best route for firefighters to take to get to the emergency?
   A. East on Bledsoe and then north on Bloomington to Brashem Avenue.
   B. East on Bledsoe, north on Halverson, and then east on Brashem Avenue to Bloomington.
   C. North on Bledsoe, east on 5th Avenue, and then south on Bloomington to Brashem Avenue.
   D. East on Bledsoe, north on Bloomington, west on Lexington, north on Halverson and then east on Brashem Avenue to Bloomington.

10. Where is Ace Industries in relation to the nearest fire station?
    A. Southwest   B. North   C. Northeast   D. East

11. An employee of Ace Industries reported smoke coming from beneath the hood of a parked car in the Ace's parking lot. What would be the best approach for firefighters responding to this emergency?
    A. East on Bledsoe, north on Bloomington, and then east on Brashem Avenue to the parking lot.
    B. North on Bledsoe, and then east on 5th Avenue to the front of Ace Industries.
    C. North on Bledsoe and then east on Brashem half a block past Bloomington.
    D. North on Bledsoe, east on Lexington, and then north on Bloomington to 5th Avenue.

12. With reference to question 11, which fire hydrant would be most appropriate to use because of its proximity to Ace Industries?
    A. Fire hydrant located at the corner of 5th Avenue and Bloomington.
    B. Fire hydrant located at the corner of Bloomington and Brashem Avenue.
    C. Standpipe connection located at the southeast corner of the Ace Industries building.
    D. Fire hydrant located at the corner of Lexington and Bloomington.

DIAGRAM C

DIRECTIONAL ORIENTATION

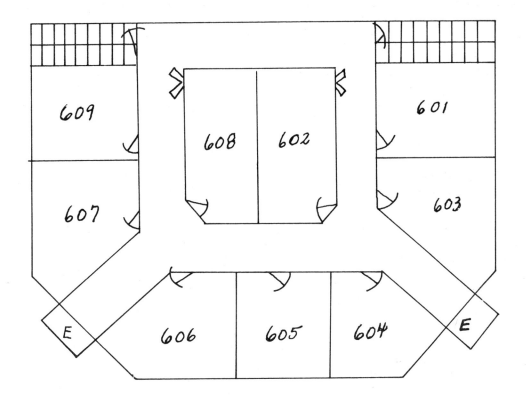

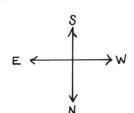

This is the basic floor plan of a six-story apartment complex.

E   represents elevators

▭   represents enclosed emergency stairwells

⋈   represents Class II standpipe hose connections

↲   represents doorways

Answer questions 13 through 15 on the basis of Diagram C

13. Apartment 601 is on fire and smoke is just beginning to seep into the hallway. According to the floor plan, which exit should be sought by the tenant in Room 604?
    A. Northwest elevator
    B. Northeast elevator
    C. Southwest emergency stairwell
    D. Southeast emergency stairwell

14. If apartment 405 were on fire, which standpipe hose connection would probably be used first by firefighters to control the blaze?
    A. Hose connection in the west hallway
    B. Hose connection in the north hallway
    C. Hose connection in the east hallway
    D. Either of the hose connections in the east and west hallways.

15. If the tenant in apartment 508 had a kitchen fire that was out of control, which apartments would be in the most immediate danger from the fire?
    A. The apartments east and west of apartment 508.
    B. The apartment north of apartment 508, and the apartment directly below (apartment 408).
    C. The apartment west of apartment 508 and the apartment directly above (apartment 608).
    D. Apartments 509, 507, and 502.

DIRECTIONAL ORIENTATION

DIAGRAM D

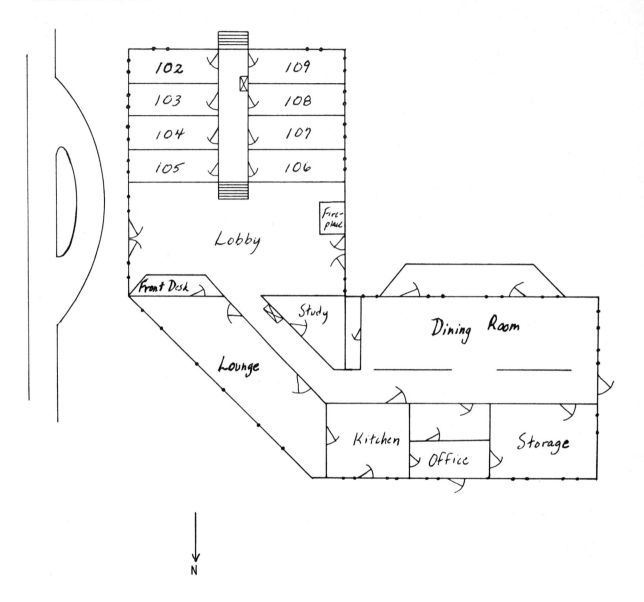

represents windows

represents stairs

represents doors

represents Class II standpipe hose connections

93

Answer questions 16 through 20 on the basis of Diagram D

16. Suppose a firefighter entered the premises via the main entrance and was told by his superior to check out the southeast section of the building. What room or rooms would that include?
    A. The kitchen and dining room
    B. Rooms 102, 103, and 104
    C. Study
    D. Lounge

17. Where is the motel's office in relation to the front desk?
    A. Southwest   B. Southeast   C. Northeast   D. Northwest

18. If a guest in room 109 noticed smoke coming from a window in the northwest section of the motel, in which room can we assume there is a fire?
    A. Room 106   B. Study   C. Kitchen   D. Dining room

19. What would be the proper sequence of directions someone would have to take to get from the west side of the dining room to room 102?
    A. East, northwest, north, and west
    B. Northeast, southeast, south, and east
    C. West, northwest, east, and west
    D. South, southeast, west, and northwest

20. If a spark from the fireplace ignited the carpeting in the lobby, what would be the appropriate exit for the tenant in room 103?
    A. North down the hallway to the main entrance
    B. West down the hallway to the main entrance
    C. South down the hallway to the side entrance
    D. Southwest down the hallway to the front desk

DIRECTIONAL ORIENTATION

DIAGRAM E

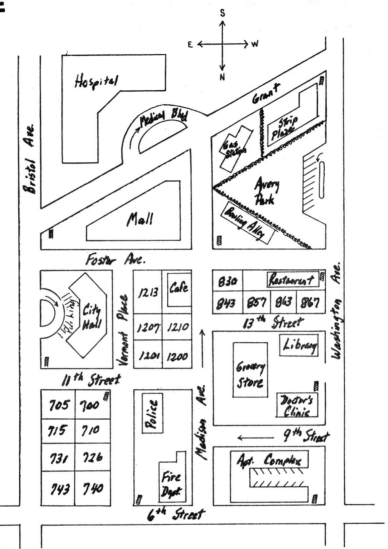

⟵ arrows indicate one way streets

▮ represent fire hydrants

⁓⁓⁓ represent fencing

addresses indicate single-family detached dwellings

Chronological sequence of events that occur during the day
1. 8:15 a.m.—a two-car collision at the corner of 13th and Madison obstructs the intersection for 50 minutes before it is cleared.
2. 10:45 a.m.—a power line falls across Foster Avenue between Vermont Place and Madison Avenue. It will take a line crew 1 1/2 hours to complete repairs.
3. 1:30 p.m.—a truck driver transporting anhydrous ammonia experiences a minor valve leak while parked at the cafe. As a safety precaution, firefighters cordon off the block while work is done to stop the leak. The incident lasts two hours.
4. 4:15 p.m.—a tractor-trailer stalls at the corner of Foster Avenue and Bristol Avenue, impeding traffic for 30 minutes before it is towed away.

95

COMPLETE FIRE-FIGHTERS EXAM PREPARATION BOOK

Answer questions 21 through 29 on the basis of Diagram E

21. At 9:00 a.m., a car was reported to be on fire at the gas station located at the intersection of Grant and Madison Avenues. What would be the best approach for firefighters responding to the emergency?
    A. South on Madison Avenue to Grant Street.
    B. North on Madison Avenue to Grant Street.
    C. South on Madison Avenue, east on 11th Street, south on Vermont Place, west on Foster Avenue, and south on Madison Avenue again to Grant.
    D. West on 6th Street, north on Washington Avenue and then southeast on Grant to Madison Avenue.

22. At 3:30 p.m., a resident at 740 Vermont Place was reported to be unconscious. After paramedics were summoned to the house, what would be the most direct route from the victim's home to the nearest medical facility?
    A. East on 6th Street, south on Bristol Avenue, and then southwest on Grant to Medical Boulevard.
    B. South on Vermont Place, east on Foster Avenue, south on Bristol Avenue, then southwest on Grant to Medical Boulevard.
    C. West on 6th Street, south on Madison Avenue, and then northeast on Grant to Medical Boulevard.
    D. West on 6th Street, south on Washington Avenue, then east on 9th Street.

23. At 11:50 a.m., paramedics just returning from the hospital were at the intersection of Grant and Bristol Avenue when a report was relayed to them that a bicyclist had been struck by a car at the intersection of Washington Avenue and 13th Street and required medical assistance. Under the prevailing circumstances, what would be the quickest way to reach the accident scene?
    A. North on Bristol Avenue, west on Foster Avenue, then north on Washington Avenue to 13th Street.
    B. North on Bristol Avenue, west on 11th Street, south on Madison Avenue, and then west on 13th Street.
    C. North on Bristol Avenue, west on Foster, north on Madison Avenue, and then west on 13th Street.
    D. Southwest on Grant and then north on Washington Avenue to 13th Street.

24. What direction is the strip plaza from the intersection of Bristol Avenue and Grant?
    A. Southwest   B. West   C. Northwest   D. Southeast

25. A patron at the restaurant located at the corner of Washington Avenue and Foster called in to report seeing a fire in an adjacent building. Without further description, the patron hung up. Because of the restaurant's location and potential vantage point, firefighters can assume which of the following buildings is probably involved?
    A. Cafe   B. Library   C. Bowling alley   D. City Hall

DIRECTIONAL ORIENTATION

26. If a resident at the apartment complex on the corner of 6th Street and Washington Avenue wanted to visit a relative in the hospital at 2:30 p.m., what would be the quickest means of getting there?
    A. East on 6th Street, south on Madison Avenue, east on Foster Avenue, south on Bristol Avenue and then southwest on Grant to Medical Boulevard.
    B. South on Washington Avenue, northeast on Grant to Medical Boulevard closest to Bristol Avenue.
    C. South on Washington Avenue, east on 9th Street, south on Madison Avenue, and then northeast on Grant to Medical Boulevard.
    D. East on 6th Street, south on Bristol Avenue, and then southwest on Grant to Medical Boulevard.

27. If a city administrator wanted to take her lunch break in Avery Park, what would be the most direct route she could use to drive there?
    A. Leave City Hall's parking lot by the southeast entrance, go south on Bristol Avenue, southwest on Grant and then north on Washington Avenue to the north entrance of the park.
    B. Leave City Hall's parking lot by the northeast exit, go south on Bristol Ave, southwest on Grant, and then north on Washington Avenue to the north entrance of the park.
    C. Leave City Hall's parking lot by the northeast exit, go south on Bristol Avenue, west on Foster Avenue and then south on Washington Avenue to the south entrance of the park.
    D. Leave City Hall's parking lot by the northeast exit, go south on Bristol Avenue, southwest on Grant, and then north on Washington Avenue to the south entrance of the park.

28. Where is City Hall in relation to the police station?
    A. South   B. Southeast   C. Northwest   D. North

29. Assuming that a fire truck was dispatched from the fire department and proceeded one block east, two blocks south, three blocks west, one block north and then half a block east, what public facility would it be nearest to at that point?
    A. Library   B. Avery Park   C. Doctors' Clinic   D. Police Department

COMPLETE FIRE-FIGHTERS EXAM PREPARATION BOOK

DIAGRAM F

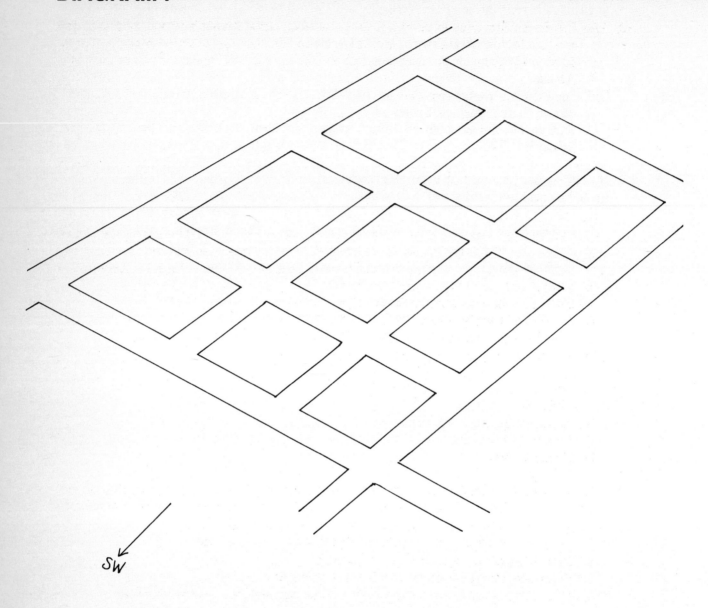

Beginning from the northwest side of this city grid the street names are Jenner Ave., Crescent St., Juniper Ave., and Boyington respectively.

Beginning from the northeast side of the same grid and going southwest, the street names are Slyvan Ln., Vandalia St., Klondyke Way, and Jefferson Blvd., respectively.

DIRECTIONAL ORIENTATION

Answer questions 30 through 35 on the basis of Diagram F

30. With the street information given, if a fire truck were to start at the north corner of this city grid and proceed one block southeast, two blocks southwest, and then two blocks southeast, at what intersection would the fire truck then be located?
    A. Klondyke Way and Boyington
    B. Vandalia Street and Juniper Avenue
    C. Jefferson Boulevard and Crescent Street
    D. Klondyke Way and Crescent Street

31. The block bounded by the streets Slyvan Lane, Juniper Avenue, Vandalia Street, and Crescent Street may be considered which block in the city grid?
    A. Northwest block   B. Northeast block   C. Southeast block   D. Central

32. A paramedic aid unit was occupied with a serious accident in the southernmost intersection shown on the city grid. If the victims involved urgently needed to be transported two blocks northeast and three blocks northwest to the closest medical facility available, we can assume that a hospital or clinic is located at the corner of what streets?
    A. Crescent Street and Vandalia Street
    B. Juniper and Klondyke Way
    C. Jenner Avenue and Vandalia Street
    D. Jenner Avenue and Slyvan Lane

33. What street is farthest from Slyvan Lane?
    A. Jefferson Boulevard
    B. Crescent Street
    C. Juniper Avenue
    D. Klondyke Way

34. Assume for the moment that Jefferson Boulevard and Vandalia Street are both one-way streets running northwest. If someone wanted to drive to the intersection of Vandalia and Boyington from the northwest end of Jefferson, what would be the quickest legal route?
    A. One block northeast, 3 blocks southeast, and then 1 block northeast
    B. Two blocks northeast, then 3 blocks southeast
    C. Double back on Jefferson to its end, then 2 blocks northeast
    D. Three blocks northeast to Sylvan, 3 blocks southeast, and then 1 block southwest.

35. In addition to the one-way streets described in the previous question, let's assume Crescent and Boyington are one-way streets as well, running in a southwesterly direction. The intersection of Slyvan and Crescent is blocked due to road construction. If an emergency vehicle needed to travel from the intersection of Jenner Avenue and Jefferson Boulevard to the intersection of Boyington and Slyvan Lane, what would be the most appropriate route?
    A. Three blocks northeast, then 3 blocks southeast
    B. Two blocks northeast, 3 blocks southeast, then 1 block northeast.
    C. One block northeast, 1 block southeast, 2 blocks northwest, then 2 blocks southeast
    D. One block northeast, 2 blocks southeast, 2 blocks northeast, then 1 block southeast

COMPLETE FIRE-FIGHTERS EXAM PREPARATION BOOK

DIAGRAM G

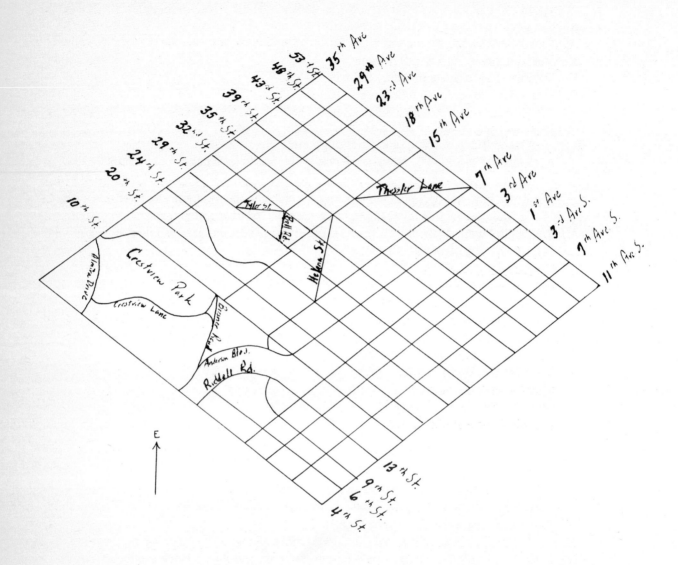

Five fire-district jurisdictions cover the area shown in the map above:

Fire Station #140 is responsible for the area bounded by 13th St., Riddell Rd., 4th St., and 11th Ave. S.

Fire Station #120 is responsible for the area bounded by 13th St. to 1st Ave., 1st Ave. to 24th St., 24th St. to 3rd Ave., 3rd Ave. to 39th St., 39th St. to 11th Ave. S, and then 11th Ave. S to 13th St.

Fire Station #125 is responsible for the area bounded by 39th St. beginning at 11th Ave. S to 3rd Ave., 3rd Ave. to 24th St., 24th St. to Helena St., Helena St. to 15th Ave., 15th Ave. to 53rd St., 53rd St. to 11th Ave. S, and then 11th Ave. S to 39th St.

Fire Station #126 is responsible for the area bounded by 15th Ave. from 53rd St. to Helena St., Helena St. to 24th St., 24th St. to 35th Ave., 35th Ave. to 53rd St., and then 53rd St. to 15th Ave.

Fire Station #130 is responsible for the area bounded by 24th St. from 35th Ave. to 1st Ave., 1st Ave. to Riddell Rd., Riddell Rd. to 4th St., 4th St. to 35th Ave., and then 35th Ave. to 24th St.

Answer questions 36 through 40 on the basis of Diagram G

36. A duplex on the north side of Helena Street would be within which fire station's jurisdiction?
    A. #126　　　　B. #125　　　　C. #130　　　　D. #120

37. The north corner of the intersection of 24th Street and 3rd Avenue would fall within which fire station's jurisdiction?
    A. #126　　　　B. #125　　　　C. #130　　　　D. #120

38. Where is the intersection of 39th Street and 3rd Avenue South in relation to the intersection of Helena Street and 29th Street?
    A. North　　　　B. South　　　　C. East　　　　D. West

39. If a vehicle were traveling northwest on 15th Avenue and went three blocks past 39th Street before making a left turn, proceeded to the next right-angled 4-way intersection, and then made a right turn to go two more blocks, at what intersection would the vehicle end up?
    A. 43rd St. and 7th Ave.
    B. 35th St. and 29th Ave.
    C. 32nd St. and 3rd Ave. S.
    D. 20th St. and 7th Ave.

40. If another vehicle were heading southwest on 10th Street from 35th Avenue and made a left turn at the first street encountered, drove another six blocks before heading south two more blocks, the vehicle would then end up at what location?
    A. Helena Street and 29th Street
    B. Tyler Street and 29th Street
    C. 39th Street and 3rd Avenue
    D. Thessler Lane and 48th Street

ANSWER SHEET FOR SAMPLE DIRECTIONAL ORIENTATION QUESTIONS

1. Ⓐ Ⓑ Ⓒ Ⓓ
2. Ⓐ Ⓑ Ⓒ Ⓓ
3. Ⓐ Ⓑ Ⓒ Ⓓ
4. Ⓐ Ⓑ Ⓒ Ⓓ
5. Ⓐ Ⓑ Ⓒ Ⓓ
6. Ⓐ Ⓑ Ⓒ Ⓓ
7. Ⓐ Ⓑ Ⓒ Ⓓ
8. Ⓐ Ⓑ Ⓒ Ⓓ
9. Ⓐ Ⓑ Ⓒ Ⓓ
10. Ⓐ Ⓑ Ⓒ Ⓓ
11. Ⓐ Ⓑ Ⓒ Ⓓ
12. Ⓐ Ⓑ Ⓒ Ⓓ
13. Ⓐ Ⓑ Ⓒ Ⓓ
14. Ⓐ Ⓑ Ⓒ Ⓓ
15. Ⓐ Ⓑ Ⓒ Ⓓ
16. Ⓐ Ⓑ Ⓒ Ⓓ
17. Ⓐ Ⓑ Ⓒ Ⓓ
18. Ⓐ Ⓑ Ⓒ Ⓓ
19. Ⓐ Ⓑ Ⓒ Ⓓ
20. Ⓐ Ⓑ Ⓒ Ⓓ
21. Ⓐ Ⓑ Ⓒ Ⓓ
22. Ⓐ Ⓑ Ⓒ Ⓓ
23. Ⓐ Ⓑ Ⓒ Ⓓ
24. Ⓐ Ⓑ Ⓒ Ⓓ
25. Ⓐ Ⓑ Ⓒ Ⓓ
26. Ⓐ Ⓑ Ⓒ Ⓓ
27. Ⓐ Ⓑ Ⓒ Ⓓ
28. Ⓐ Ⓑ Ⓒ Ⓓ
29. Ⓐ Ⓑ Ⓒ Ⓓ
30. Ⓐ Ⓑ Ⓒ Ⓓ
31. Ⓐ Ⓑ Ⓒ Ⓓ
32. Ⓐ Ⓑ Ⓒ Ⓓ
33. Ⓐ Ⓑ Ⓒ Ⓓ
34. Ⓐ Ⓑ Ⓒ Ⓓ
35. Ⓐ Ⓑ Ⓒ Ⓓ
36. Ⓐ Ⓑ Ⓒ Ⓓ
37. Ⓐ Ⓑ Ⓒ Ⓓ
38. Ⓐ Ⓑ Ⓒ Ⓓ
39. Ⓐ Ⓑ Ⓒ Ⓓ
40. Ⓐ Ⓑ Ⓒ Ⓓ

ANSWERS TO SAMPLE DIRECTIONAL ORIENTATION QUESTIONS

1. *A.* East on 10th Ave, north on Helm St, and west on 13th Ave. is the most direct route. Choice B is wrong because 12th Ave. is an eastbound one-way street. Choice C is wrong because 10th Ave. is an eastbound one-way street, too. Choice D is another possible route to the fire but choice A is the most direct.

2. *C.* The fire hydrant located at the corner of 13th Ave. and Redwood St. diagonally across from the bank is the closest.

3. *B.* The wind is blowing northeast, and the building most immediate to the bank fire in a northeast direction is the library.

4. *D.* According to the passage, the dog would end up at the corner of 10th Avenue and Helm Street. The building closest to that location is the fire station.

5. *D.* Choices A and C are not possible because of one-way streets. Choice B is a possible route, but not as short as proceeding east on 10th Avenue and then north on Redwood Street.

6. *D.* Northwest

7. *C.* North on Bledsoe and west on Brashem heads in the wrong direction. Choice B isn't correct because Brashem Ave. is an eastbound one-way street. Choice D would seem right, but the victim reportedly collapsed half a block west, not northwest, of the caretaker's residence. This places the victim closer to Brashem Avenue than 5th Avenue. The paramedics could lose valuable time crossing the park with their equipment.

8. *A.* If the wind were coming from the northeast, the property southwest of the fire would be 1208 Lexington. Choice D is not correct because all the residences shown in the diagram have successive street numbers for both Lexington and Brashem Avenue, but not Halverson.

9. *B.* All routes given could get firefighters to the emergency scene, but the fact that Bledsoe is heavily traveled by working commuters merits consideration. The time of the fire is 4:45 p.m., just about the peak traveling time for commuters going home. Therefore, it would be best to avoid as much of Bledsoe as possible. Choice B would be the proper choice.

10. *C.* Ace Industries is northeast of the only fire station shown in the diagram.

11. *A.* Choices C and D are not possible because they are counter to traffic flow. Choice B is probably the quickest way to approach Ace Industries, but the fire is in the rear of the building. Using the route in choice A would be the most direct way of approaching the smoldering car.

12. *B.* The hydrants specified in choices A and D could be utilized; however, the hydrant at the corner of Bloomington and Brashem Avenue (i.e., choice B) is the closest to the parking lot. Without a doubt, a standpipe connection in the parking lot would be ideal, but none was designated in the diagram.

13. D. Never use an elevator as a means to escape a fire. Electrical power can be interrupted, trapping people in the elevator. Leaving the premises by the southeast emergency stairwell circumvents the potential danger of walking past the apartment that is on fire.

14. D. Apartment 405 is in the same location as Apartment 605, but two levels lower. Since its position in relation to the two standpipe hose connections in the hallway is about the same, either may be used first and both may be employed depending on the severity of the blaze.

15. C. Any apartment that has a common partition with apartment 508 stands a greater chance of initially being affected than apartments separated by means of the hallway. Because heat and smoke tend to move upward, the apartment directly above 508 (i.e., 608) would be in immediate danger as well.

16. B. Rooms 102, 103, and 104 are in the southeast section of the floor plan diagrammed.

17. D. Northwest.

18. D. The vantage point afforded to a guest staying in room 109 would allow a clear view only of the dining room. Both rooms 106 and the study are at a right angle from room 109 and would be very difficult to see. The kitchen is on the other side of the building and thus impossible to view from room 109.

19. B. The correct sequence of directions for getting from the west side of the dining room to room 102 is northeast, southeast, south and then east.

20. C. The hallway outside room 103 runs north and south. This eliminates Choice B immediately. The front desk is north in relation to room 103, so choice D is incorrect. Choice A would be inappropriate because you would be running toward the fire instead of away from it.

21. C. The gas station is south on Madison Avenue at Grant, but remember the accident at 13th Street and Madison Avenue. It will not be cleared until 9:05 a.m.; therefore, just traveling south on Madison Avenue to Grant is not possible. Choice B isn't correct because it is counter to the flow of traffic on Madison Avenue. Additionally, it is heading away from the emergency. Choice D is incorrect for the same reason. A northern heading on Washington Avenue would not bring an aid unit to Grant. Instead, it is headed in the opposite direction.

22. D. Since the doctors' clinic is the closest medical facility to 740 Vermont Place, only choice D would apply. The other alternatives demonstrate various routes available to get to the hospital which is further away than the doctors' clinic on 9th Street.

23. B. Remember there is a downed power line across Foster between Vermont Place and Madison Avenue, and it will not be completely repaired by a line crew until 12:15 p.m. Therefore, choices A and C cannot be correct. Another problem with choice C is that Madison is one-way. Going north on Madison Ave. is counter to traffic flow. Choice D is not correct either, because it would require the paramedics to make a U-turn in addition to actually being the longest route to the accident scene.

24. A. Southwest.

25. C. The other alternatives do not present a clear avenue of sight from the restaurant.

26. *D.* Choices A and C are not possible because Madison is still blocked off because of the anhydrous ammonia leak from a truck parked at the cafe. Choice B is a longer route than choice D.

27. *D.* Leaving City Hall any way other than northeast is counter to traffic flow. The same applies to the park entrance. Traffic can enter only the south entrance and must depart by the north gate. With this taken into consideration, choices A and B cannot be correct. Choice C is not correct because of the fallen power line on Foster Avenue.

28. *B.* Southeast.

29. *A.* Library.

30. *A.* Klondyke Way and Boyington would be the intersection where the fire truck would end up following the directions given.

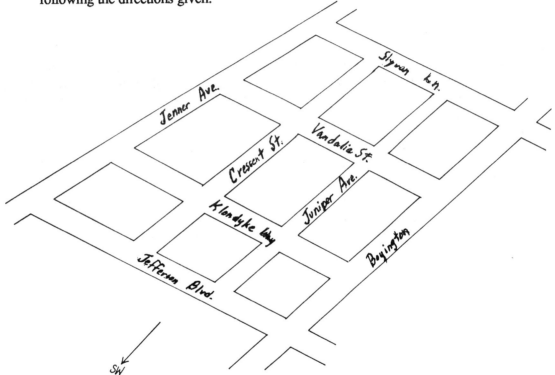

31. *B.* The northeast block. The northwest block is bounded by the streets Jenner, Vandalia, Crescent, and Klondyke Way. The southeast block is bounded by the streets Juniper, Vandalia, Boyington and Klondyke Way. The central block is bounded by Crescent, Vandalia, Juniper, and Klondyke Way.

32. *C.* Jenner Avenue and Vandalia Street. Crescent and Vandalia are two blocks northeast and two blocks northwest. Juniper and Klondyke Way is one block northeast and one block northwest. Jenner Avenue and Slyvan Lane is three blocks northeast and three blocks northwest.

33. *A.* Jefferson Boulevard. Crescent Street and Juniper Avenue both intersect Slyvan Lane. Klondyke Way runs parallel to Slyvan Lane, but two blocks away. Jefferson Boulevard parallels Slyvan Lane three blocks away.

34. A. One block northeast, three blocks southeast, and then one block northeast. Choices B and C would be counter to the one-way street direction and therefore are not correct. Choice C would also be illegal because of the doubling back. Choice D would get you to the desired location, but it is substantially longer than choice A.

35. D. One block northeast, two blocks southeast, two blocks northeast, and then one block southeast. Choices B and C run counter to the one-way street directions, and therefore can not be correct. Choice A is not possible because of the street construction taking place at the intersection of Slyvan Lane and Crescent Street.

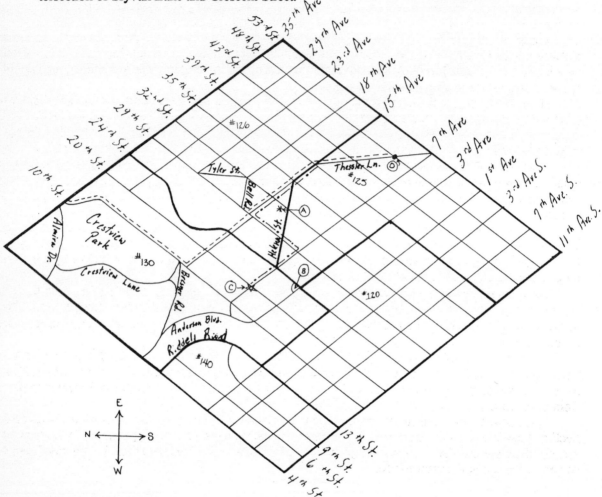

36. A. #126. (See "A" on map.)

37. C. #130. (See "B" on map.)

38. B. South.

39. D. The intersection of 20th Street and 7th Avenue. The route is marked —•—•—•— and the final destination is designated by a star on the map, lettered "C."

40. D. Thessler Lane and 48th Street. The route is marked by — — — — — and the final destination is designated as a large dot, lettered "D."

108

MATHEMATICS

Mathematics treats exact relations existing between quantities in such a way that other quantities can be deduced from them. In other words, you may know the basic quantity of a given item, but to derive further use from that quantity, it is necessary to apply known relationships (that is, formulas).

For example, let's say you wanted to know how many revolutions a tire would have to make to roll a distance of exactly 20 feet. Outside of physically rolling the tire itself and using a tape measure, it would be impossible to solve such an unknown without mathematics. However, by applying math, we can exploit known relationships to derive the answer.

If we know that the diameter of the tire is 40 inches, we can easily determine the tire's perimeter or, in other words, its circumference. In geometric terms, the tire is a circle and the known formula for determining the circumference of a circle is to multiply the diameter by π (which is 3.1416). The symbol π is referred to in mathematics as pi. Therefore, our tire's circumference is $40 \times 3.1416 = 125.66$ inches.

Since we now know that the circumference of the tire is 125.66 inches, we can learn how many revolutions a tire with this circumference would need to go exactly 20 feet. However, we cannot simply divide 125.66 inches into 20 feet because we are dealing with two entirely different units of measurement, inches and feet.

Therefore, we need to convert feet into inches. We know that there are 12 inches in 1 foot, so 20 feet $\times$ 12 inches = 240 inches. Now, we can divide 125.66 inches into 240 inches to find the answer we need. In this case, the tire would have to make 1.91 revolutions to roll exactly 20 feet. You can see by this example how known relationships can help find an unknown.

As a firefighter, you will need to have good mathematical aptitude to determine everything from volumetrics to flow dynamics, but mathematics will prove important outside your career as well. The implications can be as far reaching as calculating depreciation values on real estate for tax purposes to simply balancing your checkbook.

This section of the book is designed with the purpose of reviewing only those aspects of math that have been predominantly seen on past exams. If you find areas after completing the exercises that you have a weakness in any of these, it would be in your best interest to get additional reference material from your library.

The subjects reviewed in this section include fractions, decimals, ratios, proportions, and geometry. Each of these areas is discussed briefly, and some examples demonstrate its application. In the back of this section, there are practice exercises for you to complete. Answers and explanations are provided separately so you can check your performance.

MATHEMATICAL PRINCIPLES

A. FRACTIONS

Fractions are essentially parts of a whole. If you have ½ of something, this means you have 1 of 2 equal parts. If you have ⅞ of something, this means you have 7 of the 8 equal parts available.

The 1 of ½ is the numerator, which tells the number of parts used. The 2 is the denominator, which tells how many parts the whole has been divided into. As a general rule, if the numerator is less than the denominator, the fraction is called "proper." On the other hand, if the numerator is greater than or equal to the denominator, the fraction is called "improper." See the examples below:

⅓ is a proper fraction
⅔ is a proper fraction

$3/3$ is an improper fraction (Note: this fraction has a value of 1).
$7/3$ is an improper fraction.

A mixed number is simply a whole number with a fractional part. For example, $2\frac{1}{3}$ is a mixed number. If there is a need to change a mixed number into an improper fraction, simply multiply the whole number by the denominator of the fraction and add the resulting product to the numerator. For example:

$$2\frac{1}{3} = (2 \times 3) + 1 \text{ divided by } 3 = 7/3 \text{ an improper fraction}$$

If it is necessary to change an improper fraction into a mixed number, simply divide the numerator by the denominator. The quotient is the whole number, and the reminder is left over the denominator and the remaining fraction reduced to its lowest terms. For example:

$$\frac{15}{10} = 1\frac{5}{10} = 1\frac{1}{2}$$

The fraction $15/10$ is improper, and 15 divided by 10 is 1 with 5 left over, so $1\frac{1}{2}$ is the resulting mixed number reduced.

When we need to add, subtract, divide, or multiply fractions, certain rules need to be understood and followed. One basic rule is that multiplication or division should be done prior to addition or subtraction in mixed equations.

To start, when you add or subtract fractional numbers, you must always use a common denominator. For example:

$1/4 + 2/4 = 3/4$
$3/6 - 1/6 = 2/6$

Notice that the solutions' denominators remain the same, while the variable is the numerator (that is, $1/4 + 2/4$ does not equal $3/8$, nor does $3/6 - 1/6 = 2/0$ or 0).

The same thing applies to mixed numbers as well.

$$2\frac{1}{4} + 1\frac{3}{4} = 3\frac{4}{4}.$$

$4/4$ is an improper fraction that can be reduced to 1. Therefore,

$$2\frac{1}{4} + 1\frac{3}{4} = 4$$

But what happens when you have to add or subtract two fractions that have different denominators? Look at two such examples below:

$3/7 + 1/2 = X$
$5/8 - 1/3 = X$

Before anything can be figured out, it is essential that we find the least common denominator (LCD) for each problem. Looking at the former example ($3/7 + 1/2 = X$), we need to find the LCD for 7 and 2. In this case, it happens to be 14 (that is, 7 and 2 each divide evenly into 14, and 14 is the smallest number for which that is true). Now that we are working the problem in units of fourteenths, it is easy to figure the proportional values of the numerators involved. For example:

$$3/7 = X/14$$

To find X, you need to divide 7 into 14 and multiply the resulting quotient by the numerator:

$14 \div 7 = 2$, $2 \times 3 = 6$ therefore, $3/7 = 6/14$

Work in a similar manner for all fractions.

$1/2 = X/14$, $14 \div 2 = 7$, $7 \times 1 = 7$, therefore $1/2 = 7/14$

Now that we have a common denominator, we can add or subtract numbers as we please. In this case,

$(6/14) + (7/14) = 13/14$.

This is a proper fraction that cannot be reduced any further.

Try your hand at the latter example of $5/8 - 1/3 = X$. If you followed the format below to arrive at the answer of $7/24$, you were correct.

$5/8 = X/24$, $24 \div 8 = 3$, $3 \times 5 = 15$; therefore $5/8 = 15/24$

$1/3 = X/24$, $24 \div 3 = 8$, $8 \times 1 = 8$, therefore, $1/3 = 8/24$

$15/24 - 8/24 = 7/24$.

This is a proper fraction which cannot be reduced further.

To add or subtract mixed numbers with different fractions, the same rule applies. The only difference is that whole numbers can be treated as fractions themselves if they need to be borrowed from. See the example below:

$5\, 2/8 - 3\, 3/4 = X$

First we need to convert the fractions separately. The LCD for both fractions is 8. Therefore, $3/4 = 6/8$.

Since $2/8 - 6/8$ would leave us with a negative number, we need to borrow from the whole number (that is, 5). Therefore, we can look at $5\, 2/8$ as $4\, 10/8$. Thus, the problem now reads $4\, 10/8 - 3\, 6/8$.

As the problem now reads, we can subtract the whole numbers (4 and 3) separately, thus $4 - 3 = 1$. The fractions $10/8$ and $6/8$ can be subtracted separately as well; thus $10/8 - 6/8 = 4/8$ or $1/2$.

Now, put the whole number answer and the fractional answer together and we arrive at the total solution of $X = 1\, 1/2$.

To multiply fractions or mixed numbers it is not necessary to determine an LCD. Rather, the product of the numerators is divided by the product of the denominators. Several examples are shown below:

$$6/7 \times 5/8 = \frac{6 \times 5 = 30}{7 \times 8 = 56} = 30/56$$

which is equivalent to (or "reduces to") $15/28$

$$4 \times 7\,1/3 = 4/1 \times 22/3 = \frac{4 \times 22 = 88}{1 \times 3 = 3} = 29\,1/3$$

When you need to divide fractions or mixed numbers, convert the divisor to its reciprocal (reverse numerator and denominator) and then multiply. For example:

$7/8 \div 1/2 = X$

($2/1$ is the reciprocal of $1/2$). Thus,

$7/8 \times 2/1 = 14/8$, or $1 3/4$ reduced.

Another example involving mixed numbers is shown below:

$$6 5/8 - 3 1/3 = X$$

This equals $53/8 \times 3/10$ (reciprocal of $3 1/3$) = $159/80$, or $1 79/80$ reduced

B. DECIMALS

Decimals are basically another means to represent fractional numbers. The difference is that all fractions are expressed in factors of 10. The placement of the decimal point determines if it is a measure concerning tenths, hundredths, thousandths, ten thousandths, etc., and it will directly influence the size of the whole numbers involved. Look at the illustration below that depicts the same number with different decimal placements and examine the consequent change in value:

4459.1340 = Four thousand, four hundred fifty-nine and one hundred thirty-four thousandths.
44591.340 = Forty-four thousand, five hundred ninety-one and thirty-four hundredths.
445,913.40 = Four hundred forty-five thousand, nine hundred thirteen and four tenths.
4,459,134.0 = Four million, four hundred fifty-nine thousand, one hundred thirty-four.

When conducting addition or subtraction of decimals, the place values (that is, decimal points) of decimals must be in vertical alignment. Just as mixed numbers require a common denominator so decimals require this alignment. In this respect, the common denominator is that tenths are under tenths, hundredths are under hundredths, etc.

```
    6.5432              50.432
  + 73.43      or      - 12.07
   -------             -------
   79.9732              38.362
```

When multiplying decimals, it is necessary to treat them as whole numbers. Once you have determined the product, the decimal point is moved to the left the same number of places as there are numbers after the decimal point in both the decimals being multiplied. For example:

```
     5.678
   x  .02
   -------
     11345
     0000
   -------
    .11356
```

In this case, there are 5 numbers to the right of the decimal (678 and 02), therefore, 11356 should have the decimal placed in front of the first 1. The final number is .11356.

Dividing decimals is as simple as multiplication. When utilizing long division, simply move both place values to the right so that the divisor becomes a whole number. The decimal point then needs to be placed in the quotient above the place it has been moved to in the number being divided. At that point, each of the numbers can be treated as whole numbers and ordinary long division can be used. For example:

$$7.62 \div 3.11 = X$$
$$3.11 \overline{)7.62} = X$$

We need to move the decimal point over two places to render the divisor a whole number. Note the placement of the decimal in the quotient.

$$311.\overline{)762.} = X$$

Then,

$$\begin{array}{r} 2.450 \\ 311\overline{)762.} \\ 622 \\ \hline 1400 \\ 1244 \\ \hline 1560 \\ 1555 \\ \hline 5 \end{array}$$

and $X = 2.450$

With this rule in mind, it is very easy to convert fractions to decimals. Use the example below and observe the placement value.

The fraction $16/23$ is a proper fraction. But, when using long division to determine a decimal, we would divide 23 into 16.

$$\begin{array}{r} .6956 \\ 23\overline{)16.0} \\ 13\,8 \\ \hline 2\,20 \\ 2\,07 \\ \hline 130 \\ 115 \\ \hline 150 \\ 138 \\ \hline 12 \end{array}$$

or .696 rounded off.

C. PERCENTAGES

The term percentage by itself means divided by one hundred. For example, 15% means 15 ÷ 100. A percentage shows what portion of 100 a given number constitutes. For example, if an individual had 100 plants and gave away 20 to a friend, that would mean he or she gave away $20/100$ or .20 of the stock. To determine the percentage of plants given away, we would simply multiply .20 by 100, giving us 20%.

Let's look at another problem and determine the percentages involved.

If a fire truck had 300 feet of 1½ inch hose and three firefighters took 100 feet, 75 feet and 125 feet respectively to attend to a fire, what percentage did each firefighter carry?

Since we already know the total length of hose involved, it is a simple matter to solve for percentages.

113

	Firefighter A	Firefighter B	Firefighter C
=	$\dfrac{100 \text{ feet}}{300 \text{ feet}} \times 100$ 33%	$\dfrac{75 \text{ feet}}{300 \text{ feet}} \times 100$ 25%	$\dfrac{125 \text{ feet}}{300 \text{ feet}} \times 100$ 42%

When you total these percentages together, you get 100% of hose used.

D. RATIOS AND PROPORTIONS

A ratio is simply two items compared by division. For instance, when gears were discussed in the mechanical principles section of this book, gear ratios had implications toward mechanical advantage or speed depending on the ratio involved. (Refer to the section on gears, if necessary, for review.)

However, a proportion is an equation that shows that two ratios are equal. One of the more common types of questions seen on past exams concern speed and distance proportions. For example, if a car can travel 5 miles in 6 minutes, how far can it travel in 30 minutes, assuming the same speed is maintained? This kind of a problem would first be set up as two separate ratios and then placed in a proportion to determine the unknown.

$$\text{RATIO 1} \quad \dfrac{5 \text{ miles}}{6 \text{ minutes}} \qquad \text{RATIO 2} \quad \dfrac{X \text{ miles}}{30 \text{ minutes}}$$

In proportional form we then have; $\quad \dfrac{5 \text{ miles}}{6 \text{ minutes}} = \dfrac{X \text{ miles}}{30 \text{ minutes}}$

Once the proportion is established, you can cross multiply the proportion figures and obtain this: $6X = 5 \times 30$.

To solve for X, one of two basic algebraic laws needs to be applied. The addition law for equations states that the same value can be added or subtracted from both sides of an equation without altering the solution. The second basic law is the multiplication law for equations. This states that both sides of an equation can be multiplied or divided by the same number without changing the final solution.

These two laws are used to solve equations that have only one variable. In the case of $6X = 5 \times 30$, we will implement the multiplication/division law to determine X. If we divide both sides of the equation by 6, we can then figure how many miles the car would travel in 30 minutes.

$$\dfrac{6X}{6} = \dfrac{5 \times 30}{6}$$

$$X = \dfrac{150}{6}$$

$$X = 25 \text{ miles}$$

When working with direct proportions like this, you have to be careful not to confuse them with inverse proportions. An example would be two gears with differing numbers of teeth that run at given rpm.

Let's say one gear has 30 teeth and runs at 60 rpm, while the other gear has 20 teeth and runs at X rpm. Find X.

If we set it up as a direct proportion such as:

$$\frac{30 \text{ teeth}}{20 \text{ teeth}} = \frac{60 \text{ rpm}}{X \text{ rpm}}$$

$$\frac{30X}{30} = \frac{60 \times 20}{30}$$

$$X = \frac{1200}{30}$$

$$X = 40 \text{ rpm.}$$

Since we recall from mechanical principles that a gear with fewer teeth turns faster than a gear with more teeth, the ratios demonstrated by this proportion are incorrect. Rather, it should be inversely proportional. Therefore, it is important when coming across a question of this nature to utilize the reciprocal of one of the ratios in the equation to set up the proportion. For example:

$$\frac{20 \text{ teeth}}{30 \text{ teeth}} = \frac{60 \text{ rpm}}{X \text{ rpm}}$$

OR

$$\frac{30 \text{ teeth}}{20 \text{ teeth}} = \frac{X \text{ rpm}}{60 \text{ rpm}}$$

Both of these ways are correct proportions.

$$\frac{20X}{20} = \frac{30 \times 60}{20}$$

$$X = \frac{1800}{20}$$

$$X = 90 \text{ rpm}$$

$X = 90$ rpm is the correct answer given the fact this gear has the smaller number of teeth of the two gears.

E. GEOMETRICS

Any object that requires space has dimensions which can be measured in length, width, and height. If all three of these measurements are used to quantify the area of a given object, it can be said that it is three-dimensional or solid. If only two measurements, such as length and width, can be determined, it is considered to be two dimensional, or a plane. A line is essentially a one-dimensional figure because it has no height or width, only length.

Two-dimensional objects frequently seen in geometry are:

1. *Rectangle*—a plane formed from two pairs of parallel lines that are perpendicular to one another. Its area can be determined by multiplying length by width. For example: A rectangle measuring 9 feet by 6 feet would have an area of 54 square feet.

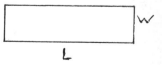

COMPLETE FIRE-FIGHTERS EXAM PREPARATION BOOK

2. *Square*—a rectangle with equal-length sides. The area for a square is found in the same way as for a rectangle.

3. *Triangle*—a closed plane shape that has three sides. Its area can be determined by multiplying ½ times the base times the height. For example: A triangle with a 10 foot base and 5 foot height has an area of 25 square feet (½ x 10 x 5). (A right triangle has one angle that is 90 degrees; that is, two sides are perpendicular).

4. *Circle*—this is a closed plane curve whose circumference is equidistant from the center. The segment of lines from the center of the circle to its circumference represents the radius. The diameter of a circle is the radius times two. The area of a circle is equal to πR^2 (π = 3.1416). For example: A circle with a radius of 10 feet has an area of $\pi \times 10^2$ = 314.16 square feet. If, on the other hand, we wanted to determine the circumference, we multiply π by the diameter or $\pi\, 2R$. In this particular case, the circumference = 3.1416 x 10 x 2 or 62.83 feet.

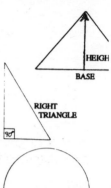

The space of a three-dimensional object in geometric terms, is called its *volume*. If we wanted to know the volume of a rectangular solid, we would take the area of a rectangle times its height. For example,

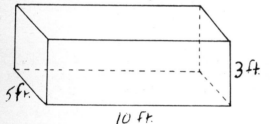

This rectangular solid has an area of 50 square feet (that is, 10 feet x 5 feet). When we multiply 50 feet x 3 feet, we can determine its volume which in this case is 150 cubic feet.

A square solid, is a cube. Since all sides are equal in length, we can simply cube the length (L^3) to determine its volume. For example,

Let's say one side measured 2 feet in length. The volume of this cube would equal 2 x 2 x 2 or 8 cubic feet.

The volume for a sphere is found by using the equation

$$V = \tfrac{4}{3}\, \pi R^3.$$

For example:

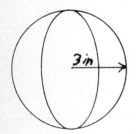

If the radius of a ball was 3 inches, what would be the ball's volume?

(4/3) x (3.1416) x (3)³ = 4/3 x 3.1416 x 27 = 113.10 cubic inches

The volume of a cylinder is found by using the equation

$$V = \pi R^2 \text{ Height}.$$

For example,

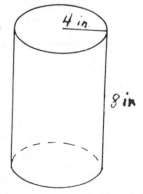

If a tin can had a radius of 4 inches and a depth of 8 inches, what is its volume?

(3.1416) x (4)² x (8) = 3.1416 x 16 x 8 = 402.12 cubic inches

One other aspect of geometry that is seen on firefighter exams concerns right triangles. The Pythagorean theorem states that the square of the side opposite the right angle equals the sum of the squares of the other sides or $A^2 + B^2 = C^2$. For example,

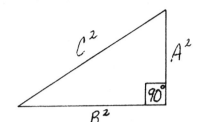

If Side A = 5 feet and Side B = 10 feet.
$5^2 + 10^2 = C^2$
$25 + 100 = C^2$
$125 = C^2$

Therefore, C is equal to the square root of 125, or 11.2.

Next you will find some sample questions on mathematical principles so you can test your skills in this area. Solutions to the problems are given at the back of the exercise to verify your work.

SAMPLE QUESTIONS ON MATHEMATICAL PRINCIPLES

1. $17\frac{3}{4} - 8\frac{1}{4} = X$. Which of the following equals X?
   A. $9\frac{2}{4}$ B. $9\frac{1}{8}$ C. $9\frac{1}{2}$ D. 9

2. $9 - \frac{3}{8} = X$. Which of the following equals X?
   A. $8\frac{3}{8}$ B. $8\frac{5}{8}$ C. 9   D. $7\frac{3}{8}$

3. $9\frac{1}{4} + 18\frac{2}{4} + 20\frac{1}{4} = X$. Which of the following equals X?
   A. $48\frac{1}{4}$ B. $48\frac{1}{2}$ C. $43\frac{1}{4}$ D. 48

4. $6\frac{1}{3} - 4\frac{5}{6} = X$. Which of the following equals X?
   A. $1\frac{1}{4}$ B. $1\frac{1}{2}$ C. $1\frac{3}{4}$ D. $1\frac{3}{6}$

5. $4\frac{2}{3} + 1\frac{1}{6} - 2\frac{1}{8} = X$. Which of the following equals X?
   A. $3\frac{17}{24}$ B. $3\frac{1}{4}$ C. $3\frac{3}{16}$ D. $3\frac{5}{18}$

6. $12\frac{3}{8} \times 2\frac{5}{7} = X$. Which of the following equals X?
   A. $32\frac{33}{25}$ B. $32\frac{17}{18}$ C. $33\frac{33}{56}$ D. $34\frac{1}{3}$

7. $7 \times \frac{1}{2} \times \frac{3}{7} = X$. Which of the following equals X?
   A. $1\frac{3}{7}$ B. $1\frac{1}{4}$ C. $1\frac{3}{4}$ D. $1\frac{1}{2}$

8. $8\frac{3}{4}$ divided by $2\frac{1}{2} = X$. Which of the following equals X?
   A. $1\frac{7}{8}$ B. $3\frac{1}{8}$ C. $3\frac{1}{2}$ D. $4\frac{1}{10}$

9. If the number $5\frac{2}{3}$ was changed from a mixed number to a decimal, which of the following is correct, assuming it is rounded off to hundredths?
   A. 5.67   B. 5.66   C. 5.6   D. 5.7

10. 6.71 x .88 = X. Which of the following equals X?
    A. 5.0948   B. 5.887   C. 5.91   D. 5.9048

11. 132.069 − 130.69 = X. Which of the following equals X?
    A. .379   B. 1.379   C. 1.739   D. 1.793

12. 8.53 + 17.671 = X. Which of the following equals X?
    A. 16.524   B. 23.102   C. 26.201   D. 25.012

13. 15.75 divided by 4.12 = X. Which of the following equals X?
    A. 3.822   B. 3.283   C. 3.023   D. 3.803

14. 6.75 + 8.372 x 3.14 = X. Which of the following equals X?
    A. 47.48   B. 33.04   C. 37.48   D. 34.03

15. 9 x 5.2 + 18.76 = X. Which of the following equals X?
    A. .40   B. 15.75   C. 28.06   D. 2.49

16. 17 − 14.87 ÷ 2.5 + 3.61 = X. Which of the following equals X?
    A. 4.46   B. .35   C. 14.66   D. 4.64

17. If 23.6 were changed into a percentage of its relationship to the number 1, which of the following would be correct?
    A. 23.6%   B. .236%   C. 236%   D. 2,360%

18. The fraction 3/7 represents what percentage?
    A. 41.85%   B. 42.85%   C. 48.25%   D. 43.35%

19. If someone were to withdraw $237.00 from a savings account that totaled $3,000.00, what percent of money is left in the account?
    A. 92.1%   B. 83.7%   C. 94.1%   D. 89.7%

20. The number 13 is 75% of what number?
    A. 15.49   B. 16.35   C. 16.99   D. 17.33

21. If a screw has a pitch that requires it to be turned 30 times to advance it two inches, what ratio correctly reflects the relationship?
    A. 2 : 30   B. 30 : 2   C. 2 : 15   D. 15 : 1

22. According to the directions on a bottle of liquid fertilizer, it is supposed to be mixed in water at the rate of 3 tablespoons per gallon before applying to a garden. How many tablespoons of fertilizer would be required for 20 gallons of water?
    A. 20   B. 40   C. 60   D. 80

23. $\dfrac{5}{8} = \dfrac{X}{32}$ Which of the following equals X?

    A. 10   B. 20   C. 25   D. 30

MATHEMATICS

24. $$\dfrac{3/5}{1/2} = \dfrac{X}{15}$$ Which of the following equals X?

   A. 18   B. 16.5   C. 19.2   D. 16

25. What is the area of a rectangle if its dimensions are 6 feet long by 4 feet wide?
    A. 10 square feet
    B. 64 square feet
    C. 24 cubic feet
    D. 24 square feet

26. If a township is a square section of territory and one side is known to be 6 miles in length, how many square miles would the township occupy?
    A. 16 square miles
    B. 18 square miles
    C. 36 square miles
    D. 42 square miles

27. What is the circumference of a gear that has a 5⅞ inch diameter?
    A. 18.05 inches   B. 16.57 inches   C. 19.45 inches   D. 18.46 inches

28. If a triangle had a base of 8 feet and a height of 3.5 feet, what would its area be?
    A. 12 square feet
    B. 14 square feet
    C. 16 square feet
    D. 20 square feet

29. If a rectangular object is 20 feet long by 15 feet wide and has a height of 4 inches, what would be its approximate volume be?
    A. 1,200 cubic feet
    B. 1,200 square feet
    C. 99 cubic feet
    D. 99 square feet

30. If one side of a cube measures 36 inches, what is its volume?
    A. 46,000 cubic inches
    B. 46,656 square inches
    C. 46,656 cubic yards
    D. 1 cubic yard

31. If a fully inflated basketball has a diameter of 12 inches, how much volume would it occupy?
    A. 904.78 cubic inches
    B. 673.54 cubic inches
    C. 509.78 cubic inches
    D. 475 cubic inches

32. If a can has a height of 16 inches and a volume of 1256.64 cubic inches, what is its diameter?
    A. 10 inches   B. 11 inches   C. 12 inches   D. 12.5 inches

33. The Pythagorean theorem concerns what kind of geometric shape?
    A. Equilateral triangle
    B. Scalene triangle
    C. Right triangle
    D. Acute triangle

34. If the length of side A is 8 feet and the length of side C (the hypotenuse) is 12.8 feet, what is the length of side B, assuming we are working with a right triangle?

    A. 8 feet
    B. 10 feet
    C. 12 feet
    D. 12.3 feet

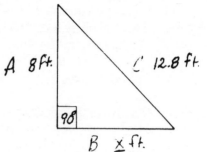

35. What is the area of the figure below?
    A. 44 square centimeters
    B. 46 square centimeters
    C. 48 square centimeters
    D. 50 square centimeters

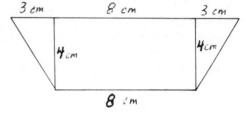

36. If a basement's floorplan had the dimensions shown below, how many square feet would this basement have?
    A. 1,000
    B. 995 square feet
    C. 988 square feet
    D. 984 square feet

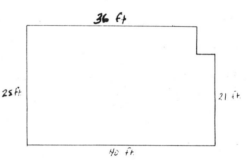

37. What is the length of a diagonal line inside a square that measures 81 square feet?
    A. 9.82 feet   B. 10.38 feet   C. 12.73 feet   D. 14.71 feet

38. If 34 inches represents 34% of the diameter of a particular circle, what is the area of the circle?
    A. 4,891 square inches
    B. 5,432 square inches
    C. 6,971 square inches
    D. 7,854 square inches

39. If you were told that a specific tire could roll 200 yards in 27.28 revolutions, what is the radius of the tire?
    A. 42 inches   B. 84 inches   C. 37.5 inches   D. 75 inches

40. What is the volume of the figure shown below? (Hint: Look at this diagram as half of a cylinder on top of a rectangular solid).
    A. 376.7 cubic centimeters
    B. 348.2 cubic centimeters
    C. 336.7 cubic centimeters
    D. 329.8 cubic centimeters

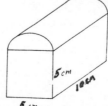

ANSWER SHEET FOR SAMPLE MATHEMATICAL PRINCIPLES QUESTIONS

1. Ⓐ Ⓑ Ⓒ Ⓓ
2. Ⓐ Ⓑ Ⓒ Ⓓ
3. Ⓐ Ⓑ Ⓒ Ⓓ
4. Ⓐ Ⓑ Ⓒ Ⓓ
5. Ⓐ Ⓑ Ⓒ Ⓓ
6. Ⓐ Ⓑ Ⓒ Ⓓ
7. Ⓐ Ⓑ Ⓒ Ⓓ
8. Ⓐ Ⓑ Ⓒ Ⓓ
9. Ⓐ Ⓑ Ⓒ Ⓓ
10. Ⓐ Ⓑ Ⓒ Ⓓ
11. Ⓐ Ⓑ Ⓒ Ⓓ
12. Ⓐ Ⓑ Ⓒ Ⓓ
13. Ⓐ Ⓑ Ⓒ Ⓓ
14. Ⓐ Ⓑ Ⓒ Ⓓ
15. Ⓐ Ⓑ Ⓒ Ⓓ
16. Ⓐ Ⓑ Ⓒ Ⓓ
17. Ⓐ Ⓑ Ⓒ Ⓓ
18. Ⓐ Ⓑ Ⓒ Ⓓ
19. Ⓐ Ⓑ Ⓒ Ⓓ
20. Ⓐ Ⓑ Ⓒ Ⓓ
21. Ⓐ Ⓑ Ⓒ Ⓓ
22. Ⓐ Ⓑ Ⓒ Ⓓ
23. Ⓐ Ⓑ Ⓒ Ⓓ
24. Ⓐ Ⓑ Ⓒ Ⓓ
25. Ⓐ Ⓑ Ⓒ Ⓓ
26. Ⓐ Ⓑ Ⓒ Ⓓ
27. Ⓐ Ⓑ Ⓒ Ⓓ
28. Ⓐ Ⓑ Ⓒ Ⓓ
29. Ⓐ Ⓑ Ⓒ Ⓓ
30. Ⓐ Ⓑ Ⓒ Ⓓ
31. Ⓐ Ⓑ Ⓒ Ⓓ
32. Ⓐ Ⓑ Ⓒ Ⓓ
33. Ⓐ Ⓑ Ⓒ Ⓓ
34. Ⓐ Ⓑ Ⓒ Ⓓ
35. Ⓐ Ⓑ Ⓒ Ⓓ
36. Ⓐ Ⓑ Ⓒ Ⓓ
37. Ⓐ Ⓑ Ⓒ Ⓓ
38. Ⓐ Ⓑ Ⓒ Ⓓ
39. Ⓐ Ⓑ Ⓒ Ⓓ
40. Ⓐ Ⓑ Ⓒ Ⓓ

MATHEMATICS

ANSWERS TO MATHEMATICAL PRINCIPLES SAMPLE QUESTIONS

1. C. $3/4 - 1/4 = 2/4$ and should be reduced to $1/2$.
   $17 - 8 = 9$, therefore, $X = 9\frac{1}{2}$

2. B. $9 - 3/8 = X$; $9 = 8\frac{8}{8}$ then, $8/8 - 3/8 = 5/8$;
   therefore, $X = 8\frac{5}{8}$

3. D. $9\frac{1}{4} + 18\frac{2}{4} + 20\frac{1}{4} = X$;
   $1/4 + 2/4 + 1/4 = 1$; $9 + 18 + 20 = 47$; therefore,
   $X = 47 + 1 = 48$

4. B. $6\frac{1}{3} - 4\frac{5}{6} = X$; $1/3 = 2/6$, therefore,
   $5\frac{8}{6} - 4\frac{5}{6} = X$, $5 - 4 = 1$ and $8/6 - 5/6 = 3/6$
   $3/6$ is reduced to $1/2$, therefore, $X = 1\frac{1}{2}$. Choice D is the correct answer also, but it is not in reduced form.

5. A. $2/3, 1/6, 1/8$ have the LCD of 24; therefore,
   $2/3 = 16/24, 1/6 = 4/24, 1/8 - 3/24$

   $$\frac{16}{24} + \frac{4}{24} = \frac{20}{24} \qquad \frac{20}{24} - \frac{3}{24} = \frac{17}{24}$$

   $(4 + 1) = 5$; $(5 - 2) = 3$; therefore $X = 3\frac{17}{24}$.

6. C. $12\frac{3}{8} = 99/8$; $2\frac{5}{7} = 19/7$
   $99/8 \times 19/7 = 1881/56 = 33\frac{33}{56}$

7. D. $\frac{7}{1} \times \frac{1}{2} = \frac{7}{2} \qquad \frac{7}{2} \times \frac{3}{7} = \frac{21}{14}$

   $21/14 = 1\frac{7}{14}$, when reduced $X = 1\frac{1}{2}$

8. C. $8\frac{3}{4} = 35/4$, $2\frac{1}{2} = 5/2$
   $35/4 \div 5/2 = 35/4 \times 2/5 = 70/20$
   $70/20 = 3\ 10/20$ or $3\frac{1}{2}$ when reduced; $X = 3\frac{1}{2}$

9. A.   The whole number 5 remains unchanged; however, $2/3$ is the same as saying 2 divided by 3. m Therefore, when rounded off to hundredths, the fraction is 0.67. Thus, the decimal should be 5.67. Choice B is correct, except that it has not been rounded off as requested. Choices C and D are not correct because both are rounded off to tenths, not hundredths.

10. D. $6.71 \times .88 = X$. $X = 5.9048$

11. B.
    132.069
 − 130.690
      1.379

therefore, $X = 1.379$

12. C.
    17.671
 + 8.53
    26.201

therefore, $X = 26.201$

13. A. 15.75 divided by 4.12 should be looked at as 1575 divided by 412, then the decimals are reinserted. The answer is 3.8223 or 3.822 rounded off to thousandths.

14. B. You should remember from high school math that multiplications and divisions are always carried out before additions or subtractions. Another similar rule states that when several multiplications and divisions occur together, you should do them in the order they are given. In this case, we first multiply 8.372 by 3.14, giving us 26.29. Now, add 26.29 to 6.75. Therefore, $X = 33.04$.

15. D. Multiply 9 x 5.2, then divide the result by 18.76.

16. C. Remembering the rules discussed in answer 14, division must be done first; 14.87 divided by 2.5 should be looked at as 148.7 divided by 25. Thus,

$$17 - 5.95 + 3.61 = X$$

   17.00   11.05
 − 5.95   + 3.61
   11.05   14.66

$X = 14.66$

17. D. 23.6 multiplied by 100 = 2360%.

18. B. 3.0 divided by 7 = .4285; then .4285 x 100 = 42.85%.

19. A.
   3000.00
 − 237.00
   2763.00

then, 2763.0 divided by 3000 = .921. Finally, .921 x 100 = 92.1%.

20. D. This kind of percentage problem needs to be worked as a proportion. We would get the following:

$$\frac{75}{100} = \frac{13}{X} \; ; 75X = 1300$$

MATHEMATICS

$$X = \frac{1300}{75} \quad 17.33$$

21. *D.* Since the ratio is ³⁰⁄₂ as with fractions, it should be reduced as low as possible, preferably to 1. In this case, 30:2 can be reduced to 15:1.

22. *C.* The ratio is 3:1 in this problem. Therefore,

$$\frac{3 \text{ Tbsp}}{1 \text{ gal}} = \frac{X \text{ Tbsp}}{20 \text{ gal}}$$

$X = 3 \times 20$ which is 60.

23. *B.*
$$\frac{5}{8} = \frac{X}{32}$$

$$8X = 160,$$

$$X = \frac{160}{8} = 20.$$

24. *A.*
$$\frac{3/5}{1/2} = \frac{X}{15} \; ; \; \tfrac{1}{2}X = 9; \text{ thus } \tfrac{1}{2}X \times 2 = 9 \times 2$$

Therefore, $X = 18$

25. *D.* A rectangular area as determined by the following equation

$A = \text{length} \times \text{width}.$

Therefore, 6 feet x 4 feet = 24 sq feet.

26. *C.* Since the area of a square is equal to the length of one side squared, we simply square 6 miles (that is, 6^2) giving us 36. Therefore, a township occupies 36 square miles.

27. *D.* The first step in this problem is to change the fraction ⅞ into a decimal since π is in decimal form. Thus, we divide 7 by 8 giving us .875. The gear's diameter is 5.875 in decimal form. The equation, Circumference = Diameter x π (3.1416). Therefore,

5.875 diameter x 3.1416 = 18.4569 inches or 18.46 inches rounded to hundredths.

28. *B.* Since the area of a triangle is equal to ½ x base x height, we can plug in the numbers accordingly giving us the following:

(½)(8)(3.5) = 14 square feet
The area is 14 square feet.

29. C. The volume of a rectangle is found by using the formula A = Length x Width x Height. Before we use this equation, all units must be the same (that is, inches or feet). In this case, it is easier to convert the height to feet.

$$\frac{4 \text{ inches}}{12 \text{ inches/feet}} = .3333 \text{ feet}$$

Area = 20 feet x 15 feet x .33 feet
Area = 99.90 cubic feet, or approximately 100 cubic feet.

30. D. The easiest way to solve this problem is to recognize that 36 inches = 1 yard. Since 36^3 is a sizable number to to multiply, we will use the simpler alternative of 1^3. Therefore, $1^3 = 1$ cubic yard. This is a common unit of measure in the construction field when ordering specific volumes of dirt, rock, concrete, etc. Choice B would be correct if the number were in cubic inches. Square inches determine only the area of a two-dimensional shape.

31. A. A basketball fully inflated can be thought of as a sphere. To determine its volume, we need to use the equation

$$\tfrac{4}{3} \times \pi (3.1416) \times R^3$$

The diameter is given as being 12 inches, therefore its radius is equal to ½ the diameter (that is, 6 inches).

$$\frac{4}{3} \times \frac{3.1416}{1} = \frac{12.5664}{3} = 4.1888$$

$6^3 = 6 \times 6 \times 6 = 216$ cubic inches
4.1888 x 216 = 904.78 cubic inches rounded to hundredths

32. A. Since the geometric shape in the question concerns a cylinder, we need to examine the equation Volume = $\pi R^2 H$. If we plug our known values into this equation, it would read

1256.64 = $\pi (3.1416)(X)^2$ (16 inches).

$$\frac{1256.642}{3.1416 \times 16} = X^2$$

25 = X^2; therefore, X = 5

Remember, 5 inches represents only the radius; the diameter would equal 5 x 2 = 10 inches.

33. C. Right triangle.

34. B. Implementing the Pythagorean theorem, $A^2 + B^2 = C^2$, we can determine X with simple algebra to solve for 1 variable.

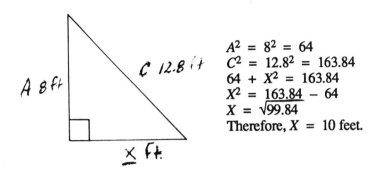

$A^2 = 8^2 = 64$
$C^2 = 12.8^2 = 163.84$
$64 + X^2 = 163.84$
$X^2 = 163.84 - 64$
$X = \sqrt{99.84}$
Therefore, $X = 10$ feet.

35. A. In geometric terms this is considered to be a trapezoid which is a quadrilateral with two sides parallel and the other two sides not parallel. To figure the area, we can see it as one rectangle (A) and two triangles (B) and (C).

The rectangular area is equal to Length x Width, therefore,

8 x 4 = 32 square centimeters (cm)

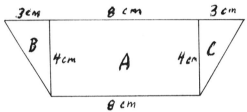

The triangle areas are equal to ½ x Base x Height

½ x 3 x 4 = 6 square cm

Since triangles B and C have the same dimension, we just multiply 6 x 2 = 12 to determine their total area combined. Therefore, $A + B + C$ = total area of trapezoid.

32 square cm + 12 square cm = 44 square cm

36. D. With the dimensions given, we can assume it has a rectangular shape. The easiest way to approach this question is to figure the total area of the basement as a rectangle and subtract the area missing in the corner.

40 feet x 25 feet = 1,000 square feet;
Side A = 25 feet − 21 feet or 4 feet
Side B = 40 feet − 36 feet or 4 feet

The area of the missing corner is 4 feet x 4 feet or 16 square feet. Therefore, this basement's total area is

1000 square feet − 16 square feet = 984 square feet.

37. C. The area of a square is the length of one side squared. If the square given is 81 square feet in area, then the square root of 81 will give us the length of the square's side which, in this case, is equal to 9. Since we are dealing with right angles in the square, we can apply the

Pythagorean theorem to determine the length of the diagonal. Therefore,

$$9^2 + 9^2 = X^2$$

$$81 + 81 = X^2$$

$$162 = X^2; \quad X = 12.73 \text{ feet}$$

38. D. First, we need to figure the diameter. If we know that 34 inches is 34% that is, .34, of the diameter, we can set up a proportion to solve it. Our proportion would be:

$$\frac{34 \text{ inches}}{.34} = \frac{X}{1.00}$$

$$\frac{.34X}{.34} = \frac{34}{.34}$$

therefore, $X = 3{,}400$ divided by 34.

Since the diameter is 100 inches, the following formula can be applied to determine the area of the circle in question.

$A = \pi R^2$. The radius is equal to ½ the diameter or in this case, 100 x .50 = 50 inches.

Therefore,

$$A = 3.1416 \times 50^2 \text{ inches}$$
$$A = 3.1416 \times 2{,}500 \text{ square inches}$$
$$A = 7{,}854 \text{ square inches.}$$

39. A. First, if we divide 27.28 revolutions into 200 yards, we can determine how many yards (or inches) this particular tire could travel after one revolution. We arrive at 7.33 yards. Then, since the diameter of this tire is referred to in inches, not yards, we simply multiply 7.33 yards x 36 inches/yard to give us 263.88 inches. In other words, for every 1 revolution this tire makes it can travel 263.88 inches. This number is the tire's circumference. If we know the tire's circumference using the equation

Diameter x 3.1416 = Circumference

we can easily figure the tires diameter.

$$X \times 3.1416 = 263.88 \text{ inches}$$

$$\frac{X \times 3.1416}{3.1416} = \frac{263.88 \text{ inches}}{3.1416}$$

$X = 83.99$ or 84 inches in diameter.

Since the question wanted the radius of the tire, we can simply divide the diameter by 2, giving us an answer of 42 inches.

40. *B.* The volume of the rectangular solid is equal to its Length x Width x Height.

   10 cm x 5 cm x 5 cm = 250 cubic centimeters.

The volume of a cylinder is equal to π x radius squared x height. Since the width of the rectangular solid can be considered the diameter of the cylinder, the radius is ½ the diameter or in this case, 5 cm divided by 2 equals the radius of 2.5 cm. The length of the rectangular solid can be considered the height of the cylinder. Therefore,

   3.1416 x 2.5^2 cm x 10 cm = volume of cylinder
   3.1416 x 6.25 cm x 10 cm = 196.35 cubic cm.

However, there is only half of a cylinder on top of the rectangular solid, so the cylinder represents only 98.2 cubic cm volume in the illustration shown. Now, add both volumes you have determined to give the total volume of this geometric shape.

   250 cubic cm + 98.2 cubic cm = 348.2 cubic cm.

READING COMPREHENSION

From the time a firefighter starts academy training to whatever level of promotion he or she achieves, a vast amount of information will have to be studied, interpreted, and applied to various situations. How efficiently that knowledge is acquired is largely dependent upon an individual's reading comprehension abilities. Some people find it easier than others to comprehend written material. Some people attribute this to inherited ability, but for the most part it is directly related to the kind of reading habits acquired in basic education. Of course, if those habits serve to impede rather than enhance a person's ability to read, it is a safe assumption that comprehension will suffer as well.

Let's examine some reading habits that were and still are indoctrinated in many English and reading classes. One such misconception is the belief that subvocalizing what you read (moving your lips or other parts of your mouth or throat as you read silently) is detrimental to your comprehension. Some teachers have even gone to the extent of passing out candy in class to prevent students from subvocalizing. In truth, subvocalizing has been proven to be beneficial in various studies. The groups of students that did subvocalize their reading were shown to have a better understanding of most material studied. This was especially true when difficult or technical information was read.

Another widespread fallacy is that students should not read word for word. Rather, it is suggested that reading should be done by looking at several words as a unit. Some say that these units lend sufficient insight into the article's content. The idea is that time is saved and reading comprehension improved. Studies demonstrate that the opposite is true.

Some teachers also believe that if a student does not fully understand the material presented, it is better to continue on instead of rereading. The line of reasoning here is that if a student does not learn what is read the first time, repetitive reading only proves to be unproductive and a waste of time. Many studies have disproved this belief. In fact, rereading can be a necessity when the material being studied is complicated or abstract. Articles should be reread as many times as necessary to get the full meaning of the text before continuing.

Another misguided belief is that any text can be completely and quickly comprehended if key words are discerned. This very concept has given rise to the speed reading industry. Speed reading "experts" claim that reading at 250 to 300 words per minute is too slow when it is entirely possible to skim at a rate of three to six times that. What they fail to mention is that comprehension is sacrificed for the sake of speed. This is analogous to the discussion of torque and speed in the mechanical section. What is gained in speed is sacrificed in torque, or *vice versa*. This raises the question: what gain is there if a fair share of information is not fully understood or comprehended?

Additionally, the major shortcoming of skim reading practices is that key words may be taken out of context and when viewed cumulatively may cause the reader to misconstrue the underlying meaning of the article. Verbs and prepositions linking nouns can dramatically alter the tenor of the material being studied. If they too, are not given attention as key words, a passage may be taken to mean one thing when, in fact, it actually means something entirely different. You can be assured that college students studying for the L.S.A.T, V.A.T., M.C.A.T., or other professional entrance exams do not skim their reading. Subject matter expected to be on these exams is closely scrutinized without regard to speed. There is no acceptable substitute for full and accurate reading comprehension.

So, to better hone your reading skills, practice reading as much material as possible and avoid the bad habits just discussed. You will find it is easier to do this if the articles you read are on topics of interest. Nothing will discourage reading more than a dull or boring article. Reading of any kind is beneficial. Newspapers, magazines, and fiction and nonfiction books are a few possible sources for material.

As you read an article, try to discern the underlying meaning of the content. What is it that the author is trying to say? Are there ideas or other information that support any conclusions? If so, which are the most important? In this respect, certain concepts can be prioritized. You will find, over time, that by following such an inquisition into all your reading, your comprehension and reading efficiency will improve immensely.

Another way to enhance reading efficiency is to develop a better vocabulary. Quite often, words will be encountered in your readings that may be unfamiliar to you. Don't skip over such words. Use a dictionary to discover their meanings and make a mental note of them. Some people find it easier to write each word on a small card as a reminder. As a challenge, try incorporating that particular word into your everyday language. A continuance of this practice is a viable means to build a strong vocabulary.

If a dictionary is not handy when you encounter an unfamiliar word, it is still possible to discern the meaning of the word. Look at the word and how it is used within the sentence. This should give you some clue as to its general meaning. For example:

> The restaurant patron was extremely *vexed* when the waiter accidentally spilled coffee on his lap.

Obviously, the customer would not be happy under such circumstances, so we know the word "vexed" implies a degree of dissatisfaction.

Another method that can be used to further understand or define an unfamiliar term uses basic word derivations or etymology. Word derivations can lend a partial, if not complete, meaning to a term. For example, take a look at the word *injudicious*. The first two letters, *in-* are a prefix that means "not" or "lack of." The root of the word, *-judi-*, means "judgment." The last portion of the word, *-ous*, is the suffix, and means "characterized by." Therefore, "injudicious" may be interpreted as a characterization of someone who lacks judgment.

The following etymology table has been provided for your convenience. This, in conjunction with viewing unfamiliar terms in context, will lend the best possible insight without the assistance of a dictionary.

COMMON PREFIXES

Prefix	Meaning	Example
a-	not or without	atypical—not typical
ab-	away from	abnormal—deviating from normal
ac-	to or toward	accredit—to attribute to
ad-	to or toward	adduce—to bring forward as evidence
ag-	to or toward	aggravate—to make more severe
at-	to or toward	attain—to reach to
an-	nor or without	anarchy—a society with no government
ante-	before or preceding	antenatal—referencing prior to birth
anti-	against or counter	antisocial—against being social
auto-	self or same	automatic—self acting
bene-	good or well	benevolence—an act of kindness or goodwill
bi-	two or twice	bisect—to divide into two parts
circum	around	circumscribe—draw a line around or encircle
com-, con-	together or with	combine—join conciliate—united or drawn together
contra-	against or opposite	contradict—opposed or against the truth
de-	removal from	decongestant—relieves or removes congestion
dec-	ten	decade—a ten year period
demi	half	demigod—partly divine and partly human
dis-, dys-	apart, negation, or reversal	dishonest—a lack of or negation of honesty
e-, ex-	from or out of	evoke—to draw forth or bring out
extra-	beyond	extraordinary—outside or beyond the usual order
hemi-	half	hemisphere—half of the globe

hyper-	excessive or over	hyperactive—excessively active
hypo-	beneath or under	hypodermic—something introduced under the skin
im-	not	impersonal—not personal or lacking personality
in-	not	inaccessible—not accessible
ir-	not	irrational—not having reason or understanding
inter-	among or between	interdepartmental—between departments
intra-	inside or within	intracellular—within a cell
intro-		introspective—to look within
kilo-	one thousand	kiloton—one thousand tons
mal-	bad or ill	malcontent—dissatisfied
mis-	wrong	misinterpret—to interpret wrongly
mono-	one or single	monochromatic—having only one color or hue
non-	not	nonresident—person who does not live in a particular place
ob-	against or opposed	object—declared opposition or disapproval
omni-	all	omnivore—an animal that eats all foods, either plant or flesh
per-	through or thoroughly	perennial—continuing or lasting through the year
poly-	many or much	polychromy—an artistic combination of different colors
post-	after or later	postglacial—after the glacial period
pre-	before or previous	preexamine—before an examination
pro-	before or supporting	proalliance—supportive of an alliance
re-	again, former state or position	reiterate—to do or say repeatedly
retro-	backward or return	retrogressive—moving backwards
self-	individual or personal	self-defense—act of defending oneself
semi-	half or part	semifinal—half final
sub-	below or under	submarine—reference to something underwater
super-	above or over	superficial—not penetrating the surface
tele-	distance	telegraph—an instrument used for communicating at a distance
trans-	across, over or through	transparent—lets light shine through
ultra-	beyond or excessive	ultraconservative—beyond ordinary conservatism
un-	not	unaccountable—not accountable or responsible

COMMON SUFFIXES

Suffix	Meaning	Example
-able, -ible	capacity capable of being	readable—able to read eligible—qualified to be chosen
-ac	like or pertaining to	maniac—like a mad person
-age	function or state of	mileage—length of distance in miles
-ally	in a manner that relates to	pastorally—in a manner that relates to rural life
-ance,	act or fact of	cognizance—knowledge through perception or reason
-ary	doing or pertaining to	subsidiary—serving to assist or supplement
-ant	person or thing	tyrant—a ruler that is unjustly severe
-ar	of the nature or pertaining to	nuclear—pertaining to nuclear matter or study
-ation	action	excavation—act or process of excavating
-cede, -ceed	to go or come	intercede—to go or come between succeed—to follow

READING COMPREHENSION

-cide	destroy or kill	homicide—the killing of a person by another
-cy	quality	decency—the state of being decent
-dy	condition or character	shoddy—pretentious condition or something poorly made
-ence,	act or fact of	despondence—loss of hope
-ery	doing or pertaining to	confectionery—place of making or selling candies or sweets
-er	of the nature or pertaining to	lawyer—a practitioner of law
-ful	abounding or full of	fretful—tending to fret or be irritable
-ic	like or pertaining to	artistic—having a talent in art
-ify	in a manner	magnify—to enlarge
-ious	full of	laborious—devoted to labor or requiring a lot of work
-ise	to make like	devise—to create from existing ideas
-ish	like	childish—acting like a child
-ism	act or practice of	capitalism—an economic system that revolves around private ownership
-ist	person or thing	idealist—a person who dreams
-ize	to make like	idolize—to make an idol of
-less	without	penniless—without a penny
-logy	the study of	archaeology—the study of historical cultures using artifacts of past activities
-ly	in a manner	shapely—well formed
-ment	the act of	achievement—the act of achieving
-ness	state of or quality	pettiness—being small in nature or insignificant
-or	of the nature or pertaining to	legislator—person responsible for legislative proceeds within gove
-ory	place	dormitory—building on or near a campus that provides living quar
-ship	condition or character	censorship—overseeing or excluding items that may be objectionable to those concerned
-tude	state of or result	solitude—state of being alone or apart from society
-ty	condition or character	levity—lightness in character
-y	quality or result	hefty—moderately heavy or weighty
-yze	to make like	analyze—to separate components into their constituent parts for observation

COMMON ROOTS

Root	Meaning	Example
acou	hearing	acoustical—pertaining to sound
acro	furthest or highest point	acrophobia—fear of heights
acu	needle	acupuncture—puncturing of body tissue for relief of pain
aero	air or gas	aeronautics—study of the operation of aircraft
alti	high	altitude—a position or a region at height
ambi	both	ambidextrous—capability of using both hands equally well
anter	in front	anteriod—toward the front
anthrop	human being	anthropology—science of mankind
aqueo, aqui	water bearing	aquatic—living in water
audio	hearing	audiology—science of hearing
auto	self	autocratic—ruled by a monarch with absolute rule

avi	bird/flight	aviary—large cage for confining birds
bia, bio	life	biography—written history of a person's life
bona	good	bonafide—with good faith
capit	head	capital—involving the forfeiture of the head or life
carb	carbon	carboniferous—containing or producing carbon or coal
carcin	cancer	carcinogen—substance that initiates cancer
carn	flesh	carnivorous—eating flesh
cent	a hundred	centennial—pertaining to 100 years
centro, centri	center	centrifugal—movement away from the center
cepha	head	hydrocephalus—condition caused by excess fluid in the head
chron	time	synchronize—to happen at the same time
citri	fruit	citric acid—sour tasting juice from fruits
corpori, corp	body	corporate—combined into one body
crypt	covered or hidden	cryptology—art of uncovering a hidden or coded message
culp	fault	culprit—one accused of a crime
cyclo	circular	cyclone—a storm with strong circular winds
demo	people	democracy—government ruled by the people through elections
doc	teach	doctrine—instruction or teaching
dox	opinion	paradox—a self contradictory statement that has plausibility of being truthful
duo	two	duologue—dialogue confined to two persons
dyna	power	dynamometer—device for measuring power
eco	environment	ecosystem—community or organisms interacting with the environment
embry	early	embryonic—pertaining to an embryo or the beginning of life
equi	equal	equilibrium—balance
ethn	race	ethnology—study of human race
exter	outside of	external—on the outside
flori	flower	florist—dealer in flowers
foli	leafy	defoliate—to strip a plant of its leaves
geo	earth	geophysics—the physics of the earth
geri	old age	geriatrics—division of medicine pertaining to old age
graphy	write	autograph—a person's own signature
gyro	spiral motion	gyroscope—rotating wheel that can spin on various planes
horti	garden	horticulture—science of cultivating plants
hydro	water	hydroplane—form of boat that glides over the water
hyge	health	hygiene—practice of preservation of health
hygro	wet	hygrometer—instrument used to measure moisture in the atmosphere
hypno	sleep	hypnology—science that treats sleep
ideo	idea	ideology—science of ideas
iso	equal	isotonic—having equal tones or tension
jur	swear	jury—body of persons sworn to tell the truth
lact, lacto	milk	lacteal—resembling milk
lamin	divided	laminate—bond together layers
lingui	tongue	linguistics—study of languages
litho	stone	lithography—art of putting design on stone with a greasy material and producing printed impressions
loco, locus	place	locomotion—act or power of moving from place to place
macro	large	macrocosm—the great world; the universe

man-	hand	manual—made or operated by hand
medi	middle	mediocre—average or middle quality
mega	large	megalopolis—urban area comprising several large adjoining cities
mero, meri	part or fraction of	meroblastic—partial or incomplete cleavage
micro	small or petty	microscopic—so small as to be invisible without the aid of a microscope
mini	small	miniature—an image or representation of something on a smaller scale
moto	motion	motive—prompting action
multi	many	multimillionaire—person with several million dollars
navi	naval	navigation—to direct course for a vessel on the sea or in the air
neo	new	neonatal—pertaining to the newborn
noct, nocti	night	nocturnal—occurring in the night
oct, octa	eight	octagonal—a shape having eight angles or eight sides
olig, oligo	scant or few	oligarchy—a government with is controlled by a few people
oo	an egg	oology—a branch of ornithology dealing with bird eggs
optic	vision or eye	optometry—profession of testing vision and examining eyes for disease
ortho	straight	orthodontics—dentistry dealing with correcting the teeth
pent, penta	five	pentagon—a shape having five angles or sides
phob	panic, or fear	arachniphobia—fear of spiders
phon	sound	phonograph—instrument for reproducing sound
pod	foot	podiatry—the study of and treatment of foot disorders
pseudo	false	pseudonym—fictitious name
psyche	mental	psychiatry—science of treating mental disorders
pyro	fire	pyrotechnics—art of making or using fireworks
quad	four	quadruped—animal having four feet
quint	five	quintuple—having five parts
sect	part or divide	bisect—divide into two equal parts
spiri	coiled	spirochete—spiral shaped bacteria
stasi	to stand still	hemostatic—serving to stop hemorrhage
techni	skill	technician—skilled person in a particular field
terri	to frighten	terrible—to excite terror
tetra	four	tetrahedron—a polyhedron with four faces
therm	heat	thermostat—device that automatically controls desired temperature
toxi	poison	toxicology—science concerning the effects, antidotes and detection of poisonous substances
uni	single	unilateral—involving one person, class or nation
urb	city	suburb—outlying part of a city
uro	urine	urology—science of studying the urinary tract and its diseases
verb	word	proverb—a name, person or thing that has become a byword
veri	truthful	verify—to prove truthful
vit	life	vitality—liveliness
vitri	glass or like glass	vitreous—resembling glass
vivi	alive	viviparous—the birth of living young
vol	wish	volunteer-to enter into or offer oneself freely
zo, zoi, zoo	animal	zoology—science of studying animal zoolife

Vocabulary exams at one time constituted a major part of the firefighter exam. Now, however, it is quite uncommon to see vocabulary test questions per se. Rather, some words not commonly used are incorporated into the context of reading comprehension questions. Without some understanding of what those terms mean, your comprehension of the article can be diminished.

Most reading comprehension test questions encountered on past exams ask questions that concern three things. First, what is the basic underlying theme of the passage, or what would be a suitable title or heading that summarizes the article? In most cases, this is an inferential question. In other words, you have to assimilate all the information given and select one of four possible options (A through D) which best encompasses the meaning of the article. There will not be a singular sentence taken directly out of the article to serve as a potential option. This is more of a judgment decision on your part.

Secondly, some questions may concern literal reading comprehension. In other words, questions about certain details, ideas, or facts will be asked. If the answer is not immediately apparent, it can simply involve going back to the applicable part of the reading and picking out the correct answer directly.

The last kind of questions may concern interpretation. After studying the information within an article, a comparable or hypothetical situation may be posed and it will be left to you to interpret how what you have read applies to it.

It is ironic, but you may find it easier to read the questions before reading the passage presented. It is somewhat of a backward approach to reading comprehension questions, but it will alert you to what is considered important, hence what to look for within the article, thus saving time.

Sample questions complete with answers and supporting explanations follow. The questions presented are not copies of past exams, but they do represent a good overview of what to expect on the actual exam.

SAMPLE QUESTIONS FOR READING COMPREHENSION

PASSAGE 1

Thermal energy flows from a hot material to a cooler material until an equilibrium is established (that is, until the temperatures of the two materials involved become the same). This is accomplished through one or a combination of three processes. The first process is conduction. Thermal energy flows from one material to another by direct contact of the two materials. This can be accomplished indirectly as well, if there is a thermal conducting medium involved. A second form of thermal energy flow is convection. As a gas or liquid within a defined air space is heated, expansion occurs, and thermal energy moves in an upward current, displacing cooler air. An example of convection would be a fire that originates at the base of a wall. As the fire builds in intensity by ascending, the cumulative heat trapped in the ceiling eventually ignites the ceiling. The third process of thermal energy flow is radiation. Thermal energy is comparable to light energy in the respect that both travel in waves. However, when radiated thermal energy is absorbed by an opaque, nonreflective material, the material involved becomes heated. If the thermal energy absorbed surpasses the material's ignition point, a fire occurs.

Answer questions 1 through 6 on the basis of Passage 1

1. What would be an appropriate title for this passage?
    A. Thermal Energy and Its Effects on Structures
    B. The Three Types of Heat Transfer
    C. The Physical Dynamics of Heat Transmission
    D. The Intrinsic Properties of Conduction, Convection, and Radiation

2. A two-story home has a vertical heating duct common to both floors. If a basement fire heated this duct to the point that the floor boards around the heat duct upstairs caught fire, what form of heat transmission can we surmise is responsible?
   A. Conduction
   B. Convection
   C. Radiation
   D. Combination of radiation and convection

3. The reason for the parapets between adjoining rooftops, as illustrated to the right, is to serve as a protective barrier against which form of heat transmission?
   A. A combination of convection and conduction
   B. Convection
   C. Conduction
   D. Radiation

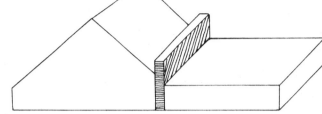

4. Every time you heat food on an electric range, you are essentially using which form of thermal energy transfer?
   A. A combination of convection and conduction
   B. Convection
   C. Conduction
   D. Radiation

5. According to the passage, which of the following is not true?
   A. Heat and light are basically the same in thermodynamic terms.
   B. As a given material is heated, it expands.
   C. When thermal energy flows from a heated source to a cooler surface, an equilibrium of temperature is reached, provided the ignition point of the material involved is not surpassed.
   D. Heated substances have the proclivity of displacing cooler substances.

6. A multi-story hotel has a centrally located open stairwell. Assume a fire starts in the mezzanine from a cigarette ash that ignited the carpeting. As the fire grows in intensity the ceiling above the stairwell suddenly exhibits flameover. What kind of thermal energy transmission can this situation mostly be attributed to?
   A. A combination of convection and conduction
   B. A combination of conduction and radiation
   C. Convection
   D. Conduction

PASSAGE 2

Fire inspection officers play a key role in preplanning fire suppression and safety considerations in a given premises. Prior to an inspection, the officer involved should be as well prepared as possible. This includes the ability to recognize potential hazards and to offer reasonable solutions. The officer must have a thorough understanding of local fire codes and how they apply to both commercial and private occupancies and have a general safety awareness with regard to how people may react during a fire. When actually performing an inspection, it is equally important to demonstrate a positive attitude that conveys a willingness to work with, not against, a property owner. Fire inspections tend to be stressful for owners.

If the owner is made aware of the fact that the officer is only trying to keep the occupants safe from hazards, there is usually better cooperation and compliance. When fire inspections are conducted in a professional manner, disagreements or arguments can, for the most part, be circumvented and appropriate safety measures implemented for the benefit of everyone concerned.

Answer questions 7 through 11 on the basis of Passage 2

7. What is probably the most important concept that can be derived from this passage?
    A. Fire inspection officers can make life miserable for owners of commercial buildings.
    B. If fire inspections are handled professionally, the intent of improving fire safety is more easily realized.
    C. It is important that all parties concerned should be made aware of the local fire codes.
    D. Fire safety officers have a rigid protocol.

8. What is probably one of the worst mistakes an inspection officer can make during an initial investigation of a new occupancy?
    A. Point out all the corrections necessary to meet full compliance with local fire codes.
    B. Dress unprofessionally.
    C. Be late to a scheduled inspection.
    D. Call attention to the enforcement policies of the fire department.

9. On the basis of maintaining a professional image, what would be the next logical step to take in the event an owner disagrees with a particular assessment?
    A. Be amiable and allow a few hazards to be overlooked.
    B. If you feel the fire codes have been adequately conveyed and the owner still refuses to comply, notify your supervisor before any further action is taken.
    C. Try to explain the reason for such fire codes.
    D. Try to dissuade the owner by pointing out how ridiculous his beliefs are.

10. Why is it imperative that an inspection be conducted with a positive attitude?
    A. Because fire inspection officers are told by their superiors to be upbeat.
    B. It advocates comraderie.
    C. It promotes a better working relationship with the property owner.
    D. The fact that someone has a positive attitude must mean that that person knows everything there is to know about fire safety.

11. According to the passage given, which of the following selections contains information not required to be known by a fire safety inspection officer?
    A. A complete familiarization with the building floor plan prior to inspection.
    B. Ability to detect possible dangers and provide sensible corrective measures.
    C. Cognizance of people's reactions during a fire.
    D. Knowledge of fire regulations and their application to various structures.

PASSAGE 3

Ventilation is an important technical concept in firefighting. Its main purpose is to reduce heat, smoke, and other dangerous gases trapped within a burning structure. The removal of these hazardous elements can be accomplished through the proper use of top level (vertical) or cross (horizontal) ventilation. Top level ventilation takes advantage of the fact that smoke and heated gases will rise to the highest point possible through convection. By providing openings in roofs and ceilings, it is possible to vent these dangerous by-products upward and outward. This provides cooler, cleaner air within the structure, making it much safer to work in. This does not mean, however, that proper breathing apparatus should not be used.

Cross ventilation uses the natural wind direction, if possible, to aid in creating a flow of air from the windward to the leeward side of a building. This is accomplished by opening windows or creating holes on opposite sides of the building. Care must be taken not to allow a fresh supply of oxygen to the fire, thereby increasing its intensity and making matters worse. Sometimes it is necessary to use large blowers and suction fans to create a crosswind if the use of a natural wind is not feasible. This is often referred to as "forced air ventilation." The main disadvantage is that it requires more manpower to implement than natural forms of ventilation.

The use of a water fog stream, a hose nozzle that creates a large volume of fine particles of water, can also be very beneficial in cooling and directing the flow of escaping heat and gases. Ventilation is a very useful technique in firefighting, but is by no means a cure-all for every problem encountered at a fire scene. No two fires are alike, and there are many factors to be considered when deciding whether or not to ventilate and what kind of ventilation to use. A few of these considerations are as follows: building type and structure, wind direction, availability of natural openings, and whether or not there is sufficient manpower available to protect surrounding areas in the event that the fire increases in intensity when the building is opened. When used properly, ventilation can be a major aid in combating a fire.

Answer questions 12 through 18 on the basis of Passage 3

12. A firefighter enters a room only to be forced back by huge columns of smoke and superheated blasts of air. He can tell that all of the smoke and heat are being trapped by the roof. According to the passage, what is the process by which all of the heat and smoke rises to the highest point of a structure?
    A. Radiation   B. Backdraft   C. Convection   D. Conduction

13. A fire breaks out in the basement of a five-story building. There are no windows in the basement. The only openings are two elevator shafts that run the height of the interior of the building. According to the passage, which type of ventilation would be best?
    A. Cross ventilation
    B. Top ventilation
    C. Water fog stream
    D. Forced air ventilation

14. The author of this passage feels that
    A. Ventilation is beneficial only on windy days.
    B. Dangerous gases are always present in a fire.
    C. Openings should always be made in a building so that air can circulate freely.
    D. Ventilation can be a great aid to firefighters when all factors are properly considered.

15. When utilizing ventilation, caution should be exercised so that
    A. As little damage is done as possible (i.e., no windows are damaged).
    B. The ventilation does not increase the fire's intensity.
    C. Smoke damage to neighboring structures is minimized.
    D. Smoke and heated air come out as many openings as possible.

16. A firefighter arrived at the fire scene at 12:02 p.m. The captain had arrived a few minutes earlier and had decided to try to use cross ventilation techniques to clear some of the huge billows of smoke coming from the house. There was a favorable wind coming from one end of the house. This firefighter should:
    A. Open windows only on the windward side of the building.
    B. Set up a windsock to make sure that he knows of any wind change.
    C. Make openings (or open windows) on the windward and leeward sides of the house.
    D. Punch four holes on each side of the house.

17. Some people, even firefighters, think that all fires are pretty much the same. This is not the case. Every time a firefighter responds to a call, he or she will be faced with new considerations and decisions. This is especially true with ventilation techniques. What works very well at one fire may be totally ineffective at another. Therefore, it is very important to take all factors into consideration each time a decision is made to ventilate. All of the following are factors that should be considered when ventilating *except*
    A. Whether the size of natural openings are within fire regulations
    B. Wind direction
    C. Building type and structure
    D. Whether enough personnel are present to protect surrounding areas

18. Forced ventilation is very useful, when no wind is present, for establishing cross ventilation. Of the following, all are advantages of this ventilation technique except
    A. Ability to control air flow
    B. Requires extra manpower to implement
    C. Helps speed smoke removal for better visibility
    D. Can be utilized in places where natural means are not feasible

PASSAGE 4

Before a fire can take place, three basic components must be present; oxygen, fuel, and heat as an ignition source. If any one or a combination of these components is missing, combustion (i.e., fire) cannot occur. In combustion, oxygen serves as an oxidizing agent and is reduced by whatever fuel may be involved, with consequent varying degrees of heat and light given off. The intensity or efficiency at which a fire burns directly depends upon the proportion or mix of the components essential for combustion. Consequently, smoke can serve as a visible indicator of that efficiency. Dark or black smoke is an indicator of incomplete combustion, a result of one or more components inadequately supplied. On the other hand, natural gas used in a kitchen stove mixes those components such that the fire which occurs burns efficiently, reducing the smoke and toxic gases (e.g., carbon monoxide) associated with incomplete combustion.

READING COMPREHENSION

Answer questions 19 through 22 on the basis of Passage 4

19. The principle idea of this passage is
    A. It is relatively easy for people to start fires.
    B. Since there are two different components involved in combustion, it would become necessary to have at least two different kinds of fire extinquishers to effectively put out a fire.
    C. A basic understanding of fire origination and behavior.
    D. How to best aid someone suffering from smoke inhalation.

20. Which of the following components is not a requirement for combustion to occur?
    A. Fuel
    B. Oxygen
    C. Carbon monoxide
    D. Heat

21. Hypothetically, if oxygen were completely depleted within the confines of a fire, as in an airtight storage locker, then
    A. The fuel involved would continue to oxidize.
    B. It would not matter because once a fire gets started, you can put it out only with water.
    C. Only partial combustion would take place.
    D. Smoke may linger; however, under those conditions, a fire cannot exist.

22. If you were faced with the task of extinquishing a fire, what method of control would prove to be most effective?
    A. Eliminate the source of the fuel.
    B. Deprive the fire of oxygen.
    C. Reduce or cool the heat (temperature) of a fire.
    D. All of the above.

ANSWER SHEET FOR SAMPLE READING COMPREHENSION QUESTIONS

1. Ⓐ Ⓑ Ⓒ Ⓓ
2. Ⓐ Ⓑ Ⓒ Ⓓ
3. Ⓐ Ⓑ Ⓒ Ⓓ
4. Ⓐ Ⓑ Ⓒ Ⓓ
5. Ⓐ Ⓑ Ⓒ Ⓓ
6. Ⓐ Ⓑ Ⓒ Ⓓ
7. Ⓐ Ⓑ Ⓒ Ⓓ
8. Ⓐ Ⓑ Ⓒ Ⓓ
9. Ⓐ Ⓑ Ⓒ Ⓓ
10. Ⓐ Ⓑ Ⓒ Ⓓ
11. Ⓐ Ⓑ Ⓒ Ⓓ
12. Ⓐ Ⓑ Ⓒ Ⓓ
13. Ⓐ Ⓑ Ⓒ Ⓓ
14. Ⓐ Ⓑ Ⓒ Ⓓ
15. Ⓐ Ⓑ Ⓒ Ⓓ
16. Ⓐ Ⓑ Ⓒ Ⓓ
17. Ⓐ Ⓑ Ⓒ Ⓓ
18. Ⓐ Ⓑ Ⓒ Ⓓ
19. Ⓐ Ⓑ Ⓒ Ⓓ
20. Ⓐ Ⓑ Ⓒ Ⓓ
21. Ⓐ Ⓑ Ⓒ Ⓓ
22. Ⓐ Ⓑ Ⓒ Ⓓ

READING COMPREHENSION

ANSWERS TO READING COMPREHENSION SAMPLE QUESTIONS

1. *C.* Choice B and D are unsatisfactory because these titles only imply discussion of the three methods of heat transmission. The basic principle of heat flow is not included by either. Choice A is incorrect because there is no discussion in the passage about heat transfers and consequent effects on structures.

2. *A.* The heating duct common to both floors of the home can serve as a thermal energy conductor. The increasing flow of heat through this medium eventually reaches an equivalent ignition point for wood. The floor boards that directly contact the heat duct consequently ignite, resulting in fire in separate quarters of the home.

3. *D.* A parapet is essentially an extension of a fire wall above the roof level. If an adjacent building's roof were on fire, the thermal energy radiated would not affect surfaces shielded behind a parapet.

4. *C.* Pots and pans serve as mediums for conducting thermal energy from the electric range heating elements to the food being cooked.

5. *A.* Heat and light were described in the passage as being similar only in the respect that they both travel in waves. These energy forms exhibit different properties because of their differences in frequency.

6. *C.* If the heated air in an open stairwell rises several floors to the ceiling and reaches a point at which ignition of the ceiling material occurs, this is due to convection. Radiation may play a small role, but for the most part it is insignificant because the distance between the mezzanine and the ceiling is too substantial.

7. *B.* If a high degree of professionalism is exhibited by fire inspection officers, it significantly enhances compliance from those parties concerned, with consequent improvements in fire safety. Choice C is true to a point but does not properly summarize what the passage is about.

8. *D.* Choice A is what needs to be done during any inspection. Laxity in this area invites complacency on the part of the owner. Both choice B and choice C are bound to detract from a professional demeanor. However, common sense dictates that choice D is the worst possible mistake. This is a place that has never been inspected before. Fire inspections can be delicate enough, without an officer demanding modification and making threats of enforcement. Remember, fire safety is achieved by working with the owner, not pitting the fire department against those involved.

9. *B.* No hazards should be overlooked in a properly conducted fire inspection, so choice A can be eliminated. Choice C may seem correct, but it is clear from the passage that it is not a fire inspection officer's place to debate the merits of local fire codes with owners; that is something to be left to the city's legislative body, in conjunction with professional counsel. Choice D is incorrect because pointing out to an owner that his or her concerns have no foundation will invariably lead to an argument. Doing what is suggested in choice B is the appropriate way of handling such a problem.

10. C. Being told by a superior to have a positive attitude is a good reason; however, what is of main concern here is the establishment of a good working relationship with the owner. Choice B is not right because the goal is not to foster a "good ole boy" attitude between the fire department and the owner. Rather it is to correct fire safety deficiencies, not become fast friends. Choice D is also wrong: an inspector may have the best possible attitude, but that doesn't assure that he or she knows all there is to know about fire safety. If something is not known by an inspector, he or she should say so up front with the assurance that an answer will be found as soon as possible. Bluffing someone with regard to any issue can lead to serious consequences.

11. A. Choices B, C, and D are necessary for a fire inspection officer to know, according to the first part of the passage. Choice A is an important factor in prefire planning; however, it is not discussed per se in the reading.

12. C. The third sentence in the reading describes convection. Choices A, B, and D were not discussed.

13. B. Under the circumstances, top level ventilation is the most viable option to firefighters. Cross ventilation isn't possible because the basement location is described as not having any windows. Forced air ventilation or water fog stream would be considered less than practical if the elevator shafts allow for ventilation. Since heat and smoke rise more quickly vertically than laterally, top level ventilation would be the most practical approach.

14. D. If ventilation is handled in an appropriate manner, fires can be more effectively controlled. Visibility within the affected structure improves, thus allowing firefighters to more easily identify hot spots and direct their lines of attack to appropriate areas. Additionally, firefighter safety improves because heat and toxic gases are purged from the building. Choice A is incorrect because ventilation is beneficial whether or not wind is present. Often, attic fires require rooftop ventilation, and in this situation wind does not play a significant role. Choice B is mentioned in the reading; however, it does not adequately present the author's intent. If choice C were to occur, the free flow of air within the affected structure would actually intensify the fire instead of aiding any suppression effort.

15. B. If firefighters utilize too much ventilation on a fire it can have the effect of fanning the flames, thus intensifying the fire instead of controlling it. This is the reason choice D is incorrect. Choice A is wrong because it is better to sacrifice a few windows or put a hole in a wall than to suffer a complete loss to fire. It is a simple matter of priorities. Choice C is not a prime concern when ventilation is being considered.

16. C. If windows are open on the windward and leeward sides of a building, effective cross ventilation will be established. Choice A is not conducive to cross ventilation. Choice B is all but a ludicrous statement. Choice D would serve to intensify the fire rather than help.

17. A. Wind direction, building type and structure, and personnel requirements were all mentioned as points to consider when ventilating. Choice A was not discussed.

18. B. The requirement of extra manpower would be considered more of a disadvantage than a help. Choices A, C, and D are all useful attributes.

19. C. The theme of this reading primarily concerns the origin and behavior of fire. Choices A and D were not discussed or implied. Choice B is incorrect because there are three components necessary for a fire to occur, not two.

READING COMPREHENSION

20. *C.* Fuel, oxygen (air), and heat are three essential ingredients for a fire to exist. On the other hand, carbon monoxide is simply a by-product of combustion.

21. *D.* If enough oxygen is not present, a fire cannot exist. Choice D is the correct answer.

22. *D.* Choices A, B, and C are all means to effectively control a fire. If any one of these components is lacking, a fire simply cannot exist.

JUDGMENT AND REASONING

This part of the exam primarily concerns the common sense of a test applicant and how he or she would react under certain situations. This may involve varied subject matter such as interdepartmental protocol or public relations, the appropriate use of equipment and related safety practices, or how best to handle certain emergency situations. This kind of exam does not necessitate complete familiarity of all aspects of firefighting. Enough information will be provided within the question to solve it solely on the basis of common sense. Keep in mind that the problem in the question must be identified first. How to best rectify that problem in the quickest and safest manner should be apparent within the scope of alternatives provided.

One other kind of question that has been seen on past exams involves tables or charts that contain specific information. On the basis of the information given, correlations or relationships need to be extrapolated, or the figures contained therein need to be understood and recognized for their significance.

The judgment and reasoning section of the exam can be very difficult to study for because there are no specific guidelines or rules to follow. Instead, test questions will draw upon wisdom acquired through everyday experiences.

Read each question completely. Often words such as *NOT, EXCEPT, LEAST, FIRST* and the like can entirely change the meaning of the question. Any potential answers that seem to be illegal or contradictory or pose a threat to any individual is probably incorrect. Options that appear self-serving or contrary to the goal of firefighters working as a team are probably wrong as well.

The test questions that follow will give you a good idea of what to expect on the actual exam. Answers are provided at the end of the exercises and give supporting reasoning for the correct choices.

SAMPLE QUESTIONS ON JUDGMENT AND REASONING

1. Below are three different hydrant outlets. Which of these would have the highest rate of discharge, or in other words, which would permit more water flow? (Arrows indicate direction of water flow.)

   A.    B.    C.    D.

2. During a search for occupants in a smoke-filled home, the probable reason for a firefighter to throw a mattress across a bed frame or form it in a U-shape is
    A. To be certain that if the mattress is on fire, it will be put out quickly by other firefighters.
    B. It is a means of helping with salvage operations.
    C. It prevents the mattress from getting soaked when water is applied.
    D. It serves to indicate that this particular room has already been searched.

3. Over a period of years, water mains usually exhibit various degrees of restricted water flow. This is most likely due to what factor?
   A. Decreased pump pressure
   B. Mineral encrustation from the water itself
   C. Toxic chemical contamination
   D. A proportional increase of air getting into the water distribution system

4. The tool shown in the diagram is most likely used for what purpose?

   A. To pick up and move energized power lines
   B. To remove lath from plaster walls or pull down ceilings
   C. To pry open locked doors or windows
   D. To pull large-diameter coiled hose out of the back of a pumper truck

5. One morning, Bill Bradshaw was late leaving for work. Immediately before leaving the house, Bill noticed his toaster starting to emit smoke. The first thing that he should have done was
   A. Leave for work anyway and worry about repairing it when he returned from work.
   B. Phone 911 to notify the fire department.
   C. Attempt to extinguish it with water.
   D. Disconnect the plug from the wall socket.

6. An eager newspaper reporter, wanting a headline story, attempts to interview a firefighter working at the scene of a two-alarm fire. When confronted by the reporter, the firefighter would best serve the department by doing what?
   A. Revealing as many details about the fire as possible.
   B. Threatening the reporter with the prospect of prosecution for criminal trespass.
   C. Enhancing public relations by emphasizing how hard firefighters work.
   D. Referring any questions the reporter may have to his superior.

Answer questions 7 through 10 on the basis of the chart below

	Flash point	Ignition temp.	Specific gravity	TLV
Phenol	185°F	1,319°F	1.058	5 ppm
Ethylene Glycol	232°F	775°F	1.115	100 ppm
Toluene	40°F	997°F	0.867	100 ppm
Napthalene	176°F	979°F	1.450	10 ppm
Crude oil	20-90°F		1.000	

Flashpoint is the minimum temperature at which a liquid in a container will give off sufficient vapor to ignite upon the application of flame.

Ignition temperature is the temperature a container must be heated to before the introduction of a flammable liquid to cause spontaneous ignition.

Specific gravity is the weight of a volume of a given substance compared to an equal volume of water. If a substance has a specific gravity of one or less, it will float on water. On the other hand, if a substance has a specific gravity greater than one, water will have a tendency to float on that substance.

TLV means *threshold limit value*, the amount of a substance that a person can inhale for five consecutive eight-hour workdays (40 hour week) without suffering any adverse effects.

7. On the basis of the information given in the chart above, which compound(s) would present an imminent flammable hazard if it inadvertently came into contact with a surface registering a temperature of 190°F?
   A. Toluene
   B. Napthalene and phenol
   C. Phenol, Toluene, napthalene and crude oil
   D. None of the above

8. Which compound demonstrates the highest toxicity to people when inhaled as a vapor?
   A. Phenol   B. Ethylene glycol   C. Toluene   D. Napthalene

9. Which chemical in the chart demonstrates a greater tendency to float in water?
   A. Phenol   B. Ethylene glycol   C. Toluene   D. Crude oil

10. From the figures given in the chart, which chemical is the most dangerous with regard to flammability?
    A. Phenol
    B. Ethylene glycol
    C. Napthalene
    D. It cannot be determined from information given.

11. Bob Gresham had just gotten off housewatch and decided to repay a fellow-co-worker for a prank. The day had been fairly slow, so Bob went to his friend's locker and filled both his workboots with water. This would be considered:
    A. Harmless because both people referred to were off-duty.
    B. Harmless because the prank only involved someone getting wet feet.
    C. Unsafe because any degree of horseplay can lead to potential injury or hinder response efficiency.
    D. Tolerable because no equipment was damaged as a result of the prank.

12. In what room of a home is a fire most likely to start?
    A. Kitchen   B. Garage   C. Living room   D. Bedroom

13. Below are four kinds of fabrics used to manufacture clothing. Which fabric would probably be the least fire resistant, assuming that all the fabrics shown are made of combustible material and are not chemically treated with fire retardant?
    A. Smooth texture, tight weave, sturdy weight
    B. Smooth texture, semitight weave, medium weight
    C. Semismooth texture, semitight weave, medium weight
    D. Fuzzy texture, loose weave, light weight

14. Some of the more expensive models of barbecue grills are fueled by liquid petroleum gas (LPG), which is contained under pressure in a steel cylinder. The fuel line must be installed correctly. If hose fittings are not tightened sufficiently, LPG may leak, creating the potential for an explosion. Which of the following methods could ascertain that hose fittings are properly installed?
    A. Apply a soapy water solution around the fittings.
    B. Run your hand the length of the hose and feel for cool spots.
    C. Smell both fittings.
    D. Listen for any hissing.

JUDGMENT AND REASONING

15. One afternoon you notice that your neighbor is unconscious in his back yard with a plugged-in electrical tool in his hand. What should you do first?
    A. Administer mouth-to-mouth resuscitation if breathing is labored or has stopped altogether
    B. Administer cardiopulmonary resuscitation (CPR) if both breathing and heartbeat have ceased.
    C. Disconnect the power to the electrical tool in your neighbor's hand.
    D. Notify the next of kin that a funeral is imminent.

16. Some types of hose couplings have a notch or groove cut into the female swivel lug and the male coupling lug. What is the most probable reason for this design?
    A. It serves to identify who the hose manufacturer is.
    B. It yields technical information with regard to how the coupling in question was made.
    C. The alignment of these markings help to prevent inadvertent cross threading.
    D. It serves to strengthen a hose coupling's structural integrity.

17. What is a major concern for firefighters while attempting to suppress a fire during the colder winter months?
    A. The hose stream actually freezes before reaching the source of the fire.
    B. Water may freeze within a hose line that is not being used.
    C. The hose lining becomes brittle and more vulnerable to rupture.
    D. Where to park a fire apparatus in relation to a fire hydrant when it is snowing.

18. If a warehouse facility recently had an automatic sprinkler system installed, which of the following qualities would not be considered advantageous?
    A. Water is discharged only in those areas affected by fire.
    B. Sprinkler heads will continue to operate regardless of the heat and smoke present.
    C. Sprinklers immediately react to fire before it develops into larger proportions.
    D. Sprinklers will continue to operate after a fire has been successfully extinguished, thus alleviating the possibility of a rekindled fire.

19. Which of the following areas would be considered the most hazardous with regard to people who smoke?
    A. A kitchen pantry with only one exit.
    B. A wooden stairway leading from the basement to the upper floor.
    C. The bed in which the smoker reads or watches television.
    D. A living room that has deep pile carpeting.

20. Firefighters often use foam in suppressing gas and oil fires. What is the probable reason this extinguishing agent is better to use over conventional water streams?
    A. Foam is cheaper.
    B. Foam has a better coolant effect on the fire than does water.
    C. Foam is lighter to carry and thus requires less effort or exertion to apply.
    D. Foam is capable of floating on the surface of a liquid fire. It serves to cool the fire and separate the fuel from its oxygen supply.

ANSWER SHEET FOR SAMPLE JUDGMENT AND REASONING QUESTIONS

1. Ⓐ Ⓑ Ⓒ Ⓓ
2. Ⓐ Ⓑ Ⓒ Ⓓ
3. Ⓐ Ⓑ Ⓒ Ⓓ
4. Ⓐ Ⓑ Ⓒ Ⓓ
5. Ⓐ Ⓑ Ⓒ Ⓓ
6. Ⓐ Ⓑ Ⓒ Ⓓ
7. Ⓐ Ⓑ Ⓒ Ⓓ

8. Ⓐ Ⓑ Ⓒ Ⓓ
9. Ⓐ Ⓑ Ⓒ Ⓓ
10. Ⓐ Ⓑ Ⓒ Ⓓ
11. Ⓐ Ⓑ Ⓒ Ⓓ
12. Ⓐ Ⓑ Ⓒ Ⓓ
13. Ⓐ Ⓑ Ⓒ Ⓓ
14. Ⓐ Ⓑ Ⓒ Ⓓ

15. Ⓐ Ⓑ Ⓒ Ⓓ
16. Ⓐ Ⓑ Ⓒ Ⓓ
17. Ⓐ Ⓑ Ⓒ Ⓓ
18. Ⓐ Ⓑ Ⓒ Ⓓ
19. Ⓐ Ⓑ Ⓒ Ⓓ
20. Ⓐ Ⓑ Ⓒ Ⓓ

ANSWERS TO SAMPLE QUESTIONS ON JUDGMENT AND REASONING

1. *C.* Obviously, the water flow in hydrant C is the least affected by friction because of rounded inner surfaces and recessed outlet connections. Surfaces of the other choices are more angular and protruding, qualities which counter flow efficiency.

2. *D.* When all view is obscured by smoke, arranging furniture in this manner makes it obvious to searchers that this particular room had already been searched. It prevents the waste of critical time when lives may be at stake.

3. *B.* Through various chemical processes, some of the mineral constituents of water tend to encrust on the internal surfaces of water mains. As the internal diameter of the main is reduced and its smooth internal surfaces transformed into an uneven and rough texture, frictional loss becomes more of a factor. The consequence is reduced water flow if the problem is left unchecked.

4. *B.* The tool diagrammed is known as a pike pole and is used to pull down ceilings and remove lath from plaster walls. The design of the tool's point should have been a clue to its intended uses. A crow bar or tool of comparable nature would be better suited to prying open locked doors or windows. Choice A is incorrect; despite the fact that pike poles can have either wooden or fiberglass handles, a firefighter can be electrocuted if an energized line comes in contact with the tool. The power should always be turned or cut off prior to any attempts to remove power lines. Choice D would never apply because the tool could potentially damage the hoses.

5. *D.* Any time an electrical appliance is involved in a fire, the source of power should immediately be disconnected. This alleviates the possibility of electrocution if choice C were attempted. Choice A would be convenient; however, there would be a good chance Mr. Bradshaw would not have a home to come back to. Choice B seems appropriate, but considering the fact that Mr. Bradshaw caught this potential fire in its earliest stages, time would be better spent disconnecting the toaster first, and then attempting to extinguish it. Contacting the fire department first may waste crucial time. It would still require a few more minutes before anyone could respond.

6. *D.* Firefighters should first and foremost worry about the emergency at hand rather than public relations. A two-alarm blaze requires the concerted effort of all firefighters on hand. If some were to conduct interviews, the efficiency of firefighting operations would diminish, thus making the blaze more difficult to bring under control. Choice A is incorrect for this reason, in addition to the fact that details released that early are incomplete at best and could hamper any criminal or arson investigation. Choice B is incorrect because there are more appropriate means to ask someone to clear an emergency scene. This is an unprofessional approach. Choice C is wrong because the fire at hand is more important than public information regarding the work ethic of your company.

7. *D.* This is something of a trick question. The key word here is *imminent*. The fact that 190°F is enough heat to cause phenol, toluene, napthalene or crude oil to vaporize does not mean it creates an imminent flammable hazard. Ignition temperatures are far above 190°F.

8. A. According to what TLV represents, if a smaller amount of a given substance is required to be inhaled before any adverse effects are felt by an individual, that substance is more toxic. Therefore, phenol is the most toxic of the substances shown in the chart.

9. C. Since toluene has the lowest specific gravity, it would exhibit a greater tendency to float on water than the other compounds given.

10. B. Since ethylene glycol has the lowest ignition temperature requirement for flammability, it can be thought of as being more dangerous than the other elements. Phenol is the most stable with regard to flammability.

11. C. Horseplay in any way, shape or form is an accident waiting to happen. It is tragic that many firefighters are injured each year as a result of horseplay. These are preventable accidents which have no place within the fire department.

12. A. Most activity within a home centers around the kitchen. Most fires occur when meals are being prepared on a stove. Temporary distractions can lead to inattentiveness. Food can easily burn within a few minutes. Oil or grease may splatter directly on the stove's heating element as well, resulting in fire. Broilers, toasters, and other electrical appliances can present a fire hazard, too, especially when unattended during use.

13. D. All of the factors mentioned in this choice make a material more flammable. Certain blends of synthetic fibers, silk, or thin cotton are much more susceptible to fire than heavier, tighter weave material. This would be comparable to trying to ignite a piece of wood and a sheet of paper. Paper will combust significantly faster than wood.

14. A. Insufficiently tightened fuel line fittings may yield no more than a trace amount of LPG once the line becomes pressurized. A small leak of this nature may not be detected by utilizing Choices B through D. Soapy water, on the hand, can make even the smallest leak evident. If tiny bubbles are being expelled around a hose fitting, that fitting will require further tightening.

15. C. Before any medical procedure is done on the victim, it is imperative that the possibility of electrocution be eliminated. Unplugging the tool would be the first step taken. At that point, medical assistance can be rendered.

16. C. This question is in reference to what is called the Higbee cut. The grooves or notches serve to indicate where the Higbee cut threading begins and tends to alleviate the problem of cross threading. The distinct advantage of having these markings is that they allow for positive identification of where the threading begins by either sight or touch. This can be of great benefit to those attempting to couple hose lengths when visibility is obscured. Choices A, B, and D are false.

17. B. An unused hose line, particularly one with a small diameter, will freeze rather quickly in sub-freezing weather, in contrast to those handlines being actively used. For all practical purposes, once a hose line is frozen it becomes useless to firefighters. Choice A is false. Choice C may be true to a limited extent, however, the hose linings manufactured these days can take significant abuse in either hot or cold conditions. Choice D would not normally be considered a major concern. If anything, it is fairly standard where a fire apparatus needs to park in relation to a fire hydrant for water supply.

JUDGMENT AND REASONING

18. *D.* Choices A, B, and C are all advantages gained by having an automatic sprinkler system installed. Choice D is a distinct disadvantage because water damage to the building and its contents need to be considered. After a fire has been extinguished, there is no longer a need to further deluge the area with more water. Firefighters can check for smoldering materials and handle them accordingly.

19. *C.* Smoking in bed poses the worst threat. As a person becomes drowsy, he or she may carelessly discard a match or drop ash on the bed linen. A fire may quickly ensue depending on the fabrics of the bedding. This, coupled with a person's slow reaction to a fire even if he or she has not already been overcome by carbon monoxide poisoning, can lead to tragic consequences. The other choices given are conditions that warrant concern; however, people are in better positions to react to these situations than to what choice C describes.

20. *D.* Choice D sums up the benefits of foam over water. Choices A and C are false or play a minor role when compared to choice D. Choice B is incorrect because water has the greatest ability to cool and absorb heat over all other compounds whether natural or man-made.

PRACTICE EXAM 1

The time allowed for the entire examination is 2 hours.

DIRECTIONS: Each question has four choices for answers, lettered A, B, C, and D. Choose the best answer and then, on the answer sheet provided, find the corresponding question number and darken with a soft pencil the circle corresponding to the answer you have selected.

PRACTICE EXAM 1

INSTRUCTIONS FOR QUESTIONS 1–15

STUDY THE FOLLOWING DESCRIPTIVE PASSAGE FOR ONLY 5 MINUTES. When your time is up, turn to the next page to answer questions 1 through 15 relating to the passage you have just studied.
YOU MAY NOT LOOK BACK AT THE PASSAGE. NOTES ARE PROHIBITED.
REMEMBER: This is a memory exercise, so any review of the passage is prohibited on the actual exam. Thus, any review on practice exams would skew your test results.

ETHYLENE OXIDE	4906610
FLAMMABLE LIQUID, CORROSIVE	UN1040
THERMALLY UNSTABLE	

    Ethylene oxide is a clear, colorless, volatile liquid with an ethereal odor. It is used to make other chemicals, and as a fumigant and industrial sterilant. It has a flash point of less than 0°F, and is flammable over a wide vapor–air concentration range. The material has to be diluted on the order of 24 to 1 with water before the liquid loses its flamability. If contaminated, it may polymerize violently with evolution of heat and rupture of its container. The vapors may burn inside a container. The vapors are irritating to the eyes, skin and respiratory system. Prolonged contact with the skin may result in delayed burns. It is lighter than water and soluble in water. The vapors are heavier than air.

If material on fire or involved in fire
    Do not extinguish fire unless flow can be stopped.
    Use water in flooding quantities as fog.
    Solid streams of water may be ineffective.
    Cool all affected containers with flooding quantities of water.
    Apply water from as far a distance as possible.
    Use "alcohol" foam, carbon dioxide or dry chemical.

If material not on fire and not involved in fire
    Keep sparks, flames, and other sources of ignition away.
    Keep material out of water sources and sewers.
    Build dikes to contain flow as necessary.
    Attempt to stop leak if without hazard.
    Use water spray to disperse vapors and dilute standing pools of liquid.

Personnel Protection
    Avoid breathing vapors.
    Keep upwind.
    Wear self-contained breathing apparatus.
    Avoid bodily contact with the material.
    Wear full protective clothing.
    Do not handle broken packages without protective equipment.
    Wash away any material which may have contacted the body with copious amounts of water or soap and water.

Evacuation
    If fire is prolonged and material is confined in the container—evacuate for a radius of 5,000 feet.
    If fire becomes uncontrollable or container is exposed to direct flame—evacuate for a radius of 1,000 feet.

From *Emergency Handling of Hazardous Materials in Surface Transportation*, Bureau of Explosives, Association of American Railroads (1981).

COMPLETE FIRE-FIGHTERS EXAM PREPARATION BOOK

ANSWER QUESTIONS 1 THROUGH 15 ON THE BASIS OF THE PASSAGE JUST READ

1. According to the passage, ethlylene oxide is a
   A. Clear, colorless, volatile gas with an evergreen odor.
   B. Brown, volatile liquid with a pungent odor.
   C. Clear, colorless, volatile liquid with an ethereal odor.
   D. Dark-colored solid that is stable to handle.

2. Ethylene oxide is used for what purpose, according to the passage?
   A. To create polymers.
   B. To make other chemicals, and as a fumigant and industrial sterilant.
   C. To produce cosmetics.
   D. To produce synthetic rubbers.

3. Where did this reading come from?
   A. Bureau of Explosives, Association of American Railroads, 1981.
   B. Bureau of Firearms, Alcohol and Tobacco.
   C. National Fire Protection Association.
   D. Hazardous Chemicals Data from the NFPA, 1976.

4. Which of the following was not suggested for personnel protection?
   A. Keep upwind of the chemical.
   B. Wear self-contained breathing apparatus.
   C. Avoid bodily contact with the chemical.
   D. Any chemical that may inadvertently contact the body should be washed off with large amounts of alcohol.

5. What was the numerical code in the upper right hand corner of the reading that apparently identified and categorized this chemical?
   A. 496610–UN 1048
   B. 4906610–UN 1040
   C. 469610–UN 1084
   D. 4610–UN 1048

6. Ethylene oxide was described as
   A. Having a flash point of 0° Fahrenheit and being flammable over a wide vapor-air concentration range.
   B. Having a flash point of 0° Celsius and being flammable over a wide vapor-air concentration range.
   C. Having a flash point greater than 0° Fahrenheit and being flammable over a narrow vapor-air concentration range.
   D. Having a flash point of less than 0° Fahrenheit and being flammable over a wide vapor-air concentration range.

7. If a container of ethylene oxide has been subjected to fire for a prolonged period of time, what is the recommended distance for safe evacuation?
   A. 1,000 feet   B. 2,000 feet   C. 3,500 feet   D. 5,000 feet

PRACTICE EXAM 1

8. What dilution of ethylene oxide with water is required to effectively inhibit its flamability?
   A. 30:1 B. 1:27 C. 24:1 D. 27:1

9. Which of the following extinguishing agents was not listed in the reading as being a viable alternative to use if ethylene oxide is on fire?
   A. Aqueous film forming foam (AFFF)
   B. Dry chemical
   C. Carbon dioxide (CO_2)
   D. Alcohol foam

10. Once ethylene oxide becomes contaminated, what kind of chemical process may occur that can generate enough heat and pressure to rupture its containment?
    A. Spontaneous combustion
    B. Polymerization
    C. Endothermic reaction
    D. Fusion

11. Which of the following statements is true, according to the reading?
    A. Ethylene oxide is heavier than water.
    B. Ethylene oxide is lighter than water.
    C. Ethylene oxide has the same density as water.
    D. Specific gravities were not discussed in the reading.

12. What improvisation was mentioned in the reading that could help contain the chemical flow if ethylene oxide is not on fire?
    A. Use an industrial vacuum cleaner.
    B. Sponge up with paper towels.
    C. Build dikes.
    D. Apply solid streams of water.

13. Which of the following choices is not recognized as a symptom stemming from ethylene oxide vapor exposure?
    A. Eye irritation
    B. Skin irritation or burns
    C. Irritation of the respiratory system
    D. Loss of muscular function

14. According to the reading, which statement below is incorrect?
    A. Solid streams of water may be ineffective in suppressing a fire involving ethylene oxide.
    B. Water sprays are an effective means of dispersing ethylene oxide vapors.
    C. Breached containers of ethylene oxide may be handled without protective equipment providing that exposure is kept to a minimum.
    D. Water should be applied to a fire involving ethylene oxide from as far away as possible.

15. Ethylene glycol vapors are what?
    A. Heavier than air
    B. Lighter than air
    C. Extremely explosive
    D. Ethylene glycol was not discussed.

16. Firefighters often apply a hose stream to the side of a building that is in close proximity to a fire. What is the primary reason for conducting such an operation?
    A. Radiant heat given off by the neighboring fire may adversely effect the nearest wall. Without the coolant effect of the hose stream, enough heat may be absorbed by an exposed surface to cause ignition and create a separate blaze.
    B. To prevent the neighboring building from getting an ash accumulation.
    C. To serve as a vapor barrier in preventing toxic gases from reaching the occupants of the neighboring building.
    D. To rinse any flammable residues from the side of the building.

17. Frequently, people use an octopus wall outlet connection to maximize the number of appliances that can be plugged into a single outlet. What would this be considered?
    A. Unsafe, because octopus electrical connections are not recommended by Underwriters Laboratories.
    B. Safe, because most electrical connections are consolidated at one point and are more easily guarded against potential trips or falls and tampering by children.
    C. Unsafe, because the excessive amount of current drawn through the outlet may exceed what it is designed to handle. This may set the stage for wires to overheat or a short circuit to occur, thus causing a fire within the partition.
    D. Safe, because if such an item can be purchased in a hardware store, it must be safe.

18. What is the greatest danger aspect for someone using an aluminum ladder to install or realign a rooftop antenna?
    A. Aluminum ladders are usually short and unstable. To compensate for lack of height, some people dangerously ascend to the topmost rung of the ladder.
    B. Aluminum is weak and more prone to collapse as compared to wooden ladders.
    C. A potential electrical shock hazard is possible when in the vicinity of power lines.
    D. Aluminum ladders are very lightweight and have a tendency to tip over.

19. In the process of fighting a fire, your superior officer directs you to perform a different part of the operation. If, because of the noise and commotion you do not fully hear or comprehend what was said, what should you do?
    A. Presume what was said was insignificant and continue with what you are doing.
    B. Ask your commanding officer to repeat the instruction.
    C. Put yourself in the shoes of the commanding officer and try to figure out on your own what he or she thinks is important.
    D. Act on your own judgment because after completion of fire academy training, you now have the skills to effectively discern what should be given priority.

20. The knot shown below is probably designed to

    A. Slip when pulled.
    B. Serve as an anchor.
    C. Alleviate load tension.
    D. Tighten up a line by taking up slack.

21. If someone's clothing were to catch on fire, what would be an appropriate thing to do?
    A. If it is a loose fitting garment, such as a jacket or coat, you should attempt to strip it off.
    B. Wrap a blanket or something similar around the victim to smother the flames.
    C. Get the victim to drop to the ground and roll slowly.
    D. All of the above are good measures.

22. Icy road conditions can be particularly hazardous for driving because
    A. Brake linings may become glazed, thus reducing the ability to come to a complete stop.
    B. Salt applied to roadway surfaces has a corrosive effect on the underbody of the vehicle.
    C. Friction between the tires and road is minimized.
    D. Wheels tend to lock up on such surfaces.

23. According to the National Safety Council, each year twenty times more deaths within the United States are caused by fires than by tornados, floods, earthquakes, and hurricanes combined. What can we deduce from this statement?
    A. Fires are more lethal by comparison.
    B. Tornados, floods, earthquakes, and hurricanes are taken for granted by most people.
    C. People are better able to cope with natural weather phenomena than fire.
    D. Statistically there are many more fires than natural weather or geophysical disasters. This, combined with public apathy and a lack of education about fire, is the probable reason for the high fatality rate.

24. When mechanical ventilation is required to help purge heat and smoke from a burning building, which of the following should be avoided?
    A. Removing windows, screens, drapes, etc., that may serve to obstruct air flow.
    B. Placing the exhaust fan, if possible, in the same direction that the prevailing winds are coming from.
    C. Using fans to bring in fresh air.
    D. Keeping the air flow path in as straight a line as possible.

25. Breaching a masonry wall is not difficult with the use of a sledge hammer. However, how a particular masonry wall is breached can make the difference between a safe entry and imminent collapse. Which of the illustrations below would probably be the safest in terms of maintaining structural integrity of the wall?

    A.   B.   C.   D.

26. What is the most important attribute a good driver has over someone with marginal driving ability?
    A. A good attitude
    B. The ability to remember every local traffic law ever written.
    C. The ability to judge distances accurately.
    D. The ability to react quickly to road hazards or other drivers who are not exercising common sense.

27. In your best judgment, what is the best reason for fire hoses to be removed and reloaded on an apparatus at least once every 30 days?
    A. Hose tends to get brittle with little use.
    B. Water trapped within a hose after use can stagnate and deteriorate the lining.
    C. It reduces the possibility of bends becoming permanently set at certain points in the hoseline.
    D. It is just another menial job to keep firefighters constantly busy.

28. Firefighters are responsible for preventing any unnecessary damage from occurring to hose lines at the scene of a fire. Not only are they expensive to replace, but damaged hoses can become ineffective in fighting a fire if they cannot remain charged because of leaks or ruptures. Which of the following hose-care practices is not recommended?
    A. Avoid excessive dragging.
    B. Don't allow vehicles to run over charged hoses.
    C. Avoid laying hose over sharp or rough corners.
    D. Use hot air to quickly dry the outside woven jacket of the hose.

29. Which of the following illustrations demonstrates proper ladder placement?

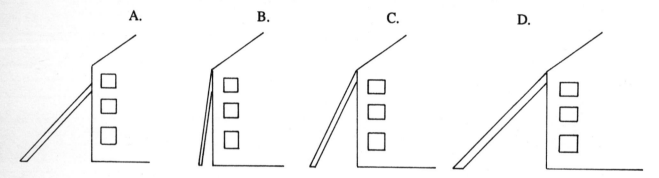

ANSWER QUESTIONS 30 THROUGH 32 ON THE BASIS OF THE CHART BELOW

Manufacturer's Recommended Safe Working Capacity for Ropes (lbs)

Diameter inches	Nylon	Manila	Dacron	Polypropylene	Braided nylon
½	726	530	492	731	1500
⅞	2145	1540	1496	1921	4740
1	2640	1800	1980	2856	5700
1½	5610	3700	3795	5525	13000
1¾	8250	5300	5610	7684	19200

30. Without considering braided nylon, which of the 1½" diameter ropes shown in the chart would support the greatest load?
    A. Braided nylon
    B. Nylon
    C. Polypropylene
    D. Manila

31. When comparing the strength factors of ⅞" diameter rope, which kind of rope would be considered the second weakest?
    A. Polypropylene B. Dacron C. Manila D. Nylon

32. According to the figures in the chart, without considering braided nylon, which rope demonstrates the second highest proportional increase in strength when its diameter is increased from ½" to 1½"?
    A. Dacron B. Manila C. Nylon D. Polypropylene

33. What is the total resistance (in OHMS) of four resistors, in a series, that are rated 7⁄16, ⅝, ½, and ¾ OHMS?
    A. 16⁄30 OHMS B. 2 5⁄16 OHMS C. 3 7⁄16 OHMS D. 8⁄15 OHMS

34. If a brick is 23⅝ centimeters in length, how many bricks laid end to end would constitute a row 283½ centimeters long?
    A. 8 bricks B. 9 bricks C. 11 bricks D. 12 bricks

35. The formula for converting Celsius temperatures to Fahrenheit is °F = 9/5 (°C) + 32. What is 25° Celsius in Fahrenheit?
    A. 102.6° B. 77° C. 67.3° D. 59.4°

36. If two engines could generate 175 and 224 horsepower, respectively, but under normal conditions exert only ⅞ and 6⁄7 of their respective ratings, what would be the expected horsepower of both engines combined?
    A. 399 horsepower
    B. 367⅕ horsepower
    C. 353 1⁄7 horsepower
    D. 345⅛ horsepower

37. If someone had an 850-liter underground diesel-fuel tank filled six times during the year, how many kiloliters of fuel were consumed annually?
    A. 5.1 kiloliters
    B. 51 kiloliters
    C. 510 kiloliters
    D. 5,100 kiloliters

38. How can 1⅓ yards to 16 inches be best expressed in terms of a ratio?
    A. 1.33 : 16 B. 1.33 : 1.5 C. 3 : 1 D. 1 : 3

39. The number 49 is 42% of what number?
    A. 20.58 B. 79.4 C. 116.67 D. 136.1

40. If Fire District #9 responded to 245 fires related to accidents or negligence, and another 8% of its fires were attributed to arson, how many calls did Fire District #9 receive altogether? (All fires can be regarded as caused by arson, negligence, or accident)
    A. 248 B. 253 C. 266 D. 275

41. If pulley A has a circumference of 14.75 inches and pulley B has a circumference of 23.25 inches, how fast would pulley A turn if pulley B were turning at 200 rpm?

    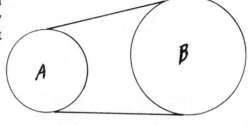

    A. 315.25 rpm B. 274.67 rpm

    C. 192.83 rpm D. 126.88 rpm

42. The loss of water pressure in a hose can be attributed to friction. If it is known that the friction loss in 100 feet of 1½ inch hose is 25 pounds per square inch, what would be the friction loss for 150 feet of 1½ inch hose, assuming all other factors remain constant?
    A. 16.67 psi B. 37.5 psi C. 47.3 psi D. 52 psi

43. If a 6-foot-tall firefighter cast a shadow 7.5 feet long and simultaneously noticed that the fire station cast a shadow 35 feet long, how tall is the fire station?
    A. 27.5 feet B. 28 feet C. 29.7 feet D. 32 feet

44. If the two pulleys shown are spaced 35 centimeters apart center to center and have 10.5 centimeter diameters, what would be the length of V-belt required?

    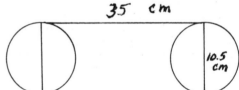

    A. 95.72 cm B. 97.691 cm

    C. 100.2 cm D. 102.987 cm

45. What would be the displacement of a piston that has a bore of 3.25 inches and a stroke of 4.75 inches?
    A. 39.41 cubic inches
    B. 47.59 cubic inches
    C. 132.52 cubic inches
    D. 157.62 cubic inches

46. Someone was told to cut a 23.5-inch-diameter circle from a square piece of sheet metal that measures 24 inches on one side. Approximately how many square inches of sheet metal would be left after the circular cut was made?
    A. 142.26 square inches
    B. 147 square inches
    C. 159.87 square inches
    D. 163.2 square inches

47. If a firefighter applicant scored 80%, 82%, 90% and 87% on his first four exams, what kind of grade would be required on the fifth exam to acquire an overall average of 87%?
    A. 87%   B. 89%   C. 96%   D. 98%

48. If six firemen can conduct waterflow analysis on 25 fire hydrants in two days, how many fire hydrants could be inspected by five firemen in four days?
    A. 22   B. 27   C. 35   D. 42

ANSWER QUESTIONS 49 THROUGH 54 ON THE BASIS OF THE PASSAGE BELOW

Since no kind of fire extinguisher is effective on all types of fires, it is imperative that the appropriate kind of extinguishing agent be used for the situation at hand. Otherwise, a fire can be made worse or the potential for explosion can be created. Therefore, it is important that an individual understand the various classifications of fire extinguishers and their appropriate uses.

There are essentially four classes of fires, which are identified as A, B, C, or D. Class A fires principally involve combustibles such as paper, wood, or cloth. Class B fires originate from flammable petroleum products, gases, or other combustible liquids. Class C fires are basically the same as Class A or Class B fires, but have an underlying potential for danger from electricity. This kind of fire necessitates an extinguishing agent that has nonconductive properties to circumvent electrical shock. One example would be the dry chemical sodium bicarbonate. The last and perhaps the most complicated fires to handle are Class D fires. This classification principally involves combustible metals and extremely volatile solids. Examples include magnesium, potassium, sodium, and zirconium. Fires of this nature require special techniques to achieve suppression and specific extinguishing agents.

More often than not a fire falls into two or more classes. Consequently, it is advantageous to have versatile fire extinguishers on hand. For example, a fire extinguisher that is rated A, B, C is better than another extinguisher rated only A, B in that the former has the additional ability to be used when electrical hazard is present. Whenever contemplating fire safety, be sure to have the right kind of fire extinguisher for the kind of conditions present. Failure to observe this concept invites tragedy.

49. Which of the choices below would best summarize the information given in this passage?
    A. How to best use a fire extinguisher.
    B. Multiple letter rating on a fire extinguisher indicates greater versatility for suppressing different kinds of fires.
    C. Using the correct extinguishing agent on a fire is very important.
    D. Each fire is unique and can be thought of as belonging to one of four classes (A, B, C, or D).

50. On the basis of what was discussed in the passage, what kind of fire extinguisher would be the most suitable for a woodscrap fire?
    A. Class A   B. Class B   C. Class C   D. Class D

51. A drill press operator noticed the metallic waste on the floor was smoking and on the verge of igniting. What kind of fire extinguisher would be best suited for this situation?
    A. Class A   B. Class B   C. Class C   D. Class D

COMPLETE FIRE-FIGHTERS EXAM PREPARATION BOOK

52. If sawdust that had accumulated on the underside of a table saw caught fire in a woodshop, what kind of fire extinguisher should be used to suppress it?
    A. Class A
    B. Class A B
    C. Class A B C
    D. Class D

53. Sodium bicarbonate is an example of what kind of extinguishing agent?
    A. Soda acid
    B. Salt water
    C. Liquified compressed gas
    D. Nonconductive dry chemical

54. If a 2½ gallon fire extinguisher contained water stored under pressure, but lacked a face plate identifying its classification, what class or classes of fires could we assume this unit would be effective in suppressing?
    A. Class A
    B. Class B C
    C. Class A B C
    D. Class D

ANSWER QUESTIONS 55 THROUGH 62 ON THE BASIS OF THE PASSAGE BELOW

Hazardous material stored in fixed facilities such as tanks or buildings require proper identification. With that information available, fire inspectors know of what they are dealing with and how to take appropriate action in the event of leaks, spills, fires, or exposure by victims. The National Fire Protection Association has developed a recognizable labeling system that addresses such issues. This is known as the 704 label. It is essentially a diamond-shaped placard which is divided into four different colored quadrants.

The left quadrant shown is blue and indicates how hazardous the material in question is to a person's health.

The top quadrant shown is red and represents the flamability of the material or, in other words, how susceptible the material is to burning.

The right quadrant is yellow and represents the reactivity of the hazardous material present. This quantifies the susceptibility of a hazardous material to release energy.

The bottom quadrant specifies whether the hazardous material stored poses a radioactive danger which is represented by the insignia or has potential reactivity with water which is represented by the letter W with a horizontal slash bisecting it.

One other feature of this labeling system is that the degree of health hazard, flammability, and reactivity posed by a hazardous material is indicated by the use of a number system. The five signals (i.e., numbers) used are 0 through 4. The number 0 is indicative of a fairly stable substance posing minimum risk to those who handle it, whereas the signal 4 is indicative of an extremely toxic or unstable material. The information afforded to an inspector by this kind of labeling system allows for a safer approach, quicker recognition of the material in question, and the ability to take appropriate measures in the event of an emergency.

55. What would be an appropriate title for this article?
    A. The Emergency Response Guide As Adopted by the Department of Transportation (DOT)
    B. Identification Placards for Transporting All Hazardous Materials
    C. The NFPA System For Identifying Hazardous Materials
    D. Fire Inspections Made Easier with the NFPA Labeling System

56. What is the name of the labeling system discussed in the article?
    A. NFPA hazardous material insignia
    B. The diamond placard hazardous materials identification system
    C. Hazardous materials color-coded quadrant system
    D. NFPA 704 labeling system

57. If an aboveground storage tank contains a hazardous material that is extremely flammable, but poses minimal risk to an individual's health, and is fairly stable, which of the placards shown below would be appropriate to it?

A. 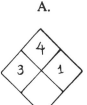 B. 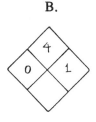 C. 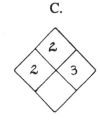 D.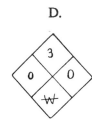

58. If another aboveground storage tank contains a radioactive material that poses an extreme hazard to those who do not wear protective gear, which of the choices below would best identify the substance in question according to the 704 labeling system?
    A. The blue quadrant must demonstrate a signal of 3 or 4 coupled with the radioactivity insignia in the bottom quadrant of the placard.
    B. The blue quadrant must demonstrate a signal of 1 or 2 in conjunction with the yellow quadrant bearing the insignia for radioactivity.
    C. The red quadrant must show a high number to indicate an extreme reactivity due to the presence of radiation.
    D. All quadrants should be marked with a high signal due to the radioactivity involved.

59. What is the color of the quadrant in a 704 label that indicates a hazardous material's potential reactivity?
    A. White   B. Blue   C. Red   D. Yellow

60. What is the color of the quadrant in a 704 label that indicates a hazardous material's inherent reactivity to water?
    A. White
    B. Blue
    C. Yellow
    D. Was not mentioned in the article

61. If firefighters respond to a chemical spill inside a warehouse and the drum from which the chemical spilled had a 704 label bearing the letter W with a slash through it, what would be the one thing that firefighters would immediately realize?
    A. It amounts to a very serious situation that calls for a complete evacuation of the premises.
    B. The hazardous material spilled is reactive with water. Therefore pressurized water should not be used to control a potential fire nor should water be used for cleanup.
    C. Self-contained breathing apparatus should be used for handling such a material.
    D. The chemical spill poses a radiation danger to those working in close proximity without the benefit of protective gear.

62. As far as fire inspectors are concerned, which of the following alternatives is not a benefit of labeling hazardous materials in stored containers?
    A. It is conducive to safer access.
    B. Corrective action can be taken without exacerbating the situation.
    C. It protracts the amount of time necessary to conduct a thorough cleanup.
    D. It allows for faster discernment of the material involved.

ANSWER QUESTIONS 63 THROUGH 67 ON THE BASIS OF THE PASSAGE BELOW.

A fuel source, be it in a solid, liquid or gaseous state, must meet certain criteria before combustion can actually take place. Fuel with a definitive shape and size is considered to be a solid. Another characteristic inherent to a solid is the direct correlation of surface area and the ability to absorb heat. The larger the surface area of a solid, the more quickly it can heat up. However, before a solid fuel can combust, it is necessary to convert the solid to a gaseous state. This is accomplished by a chemical decomposition process that involves the application of heat. Another term for this conversion is Pyrolysis.

A liquid is any substance that assumes the shape of its container. (But unlike gasses, liquids are not compressible.) Like solids, liquids require the conversion to a gaseous state to become combustible. This process is known as vaporization. The flash point of a liquid is the temperature required to vaporize a liquid. In layman's terms, any liquid that has a low flash point is particularly susceptible to igniting. Such liquids are identified as volatile. Butyl alcohol is one such example.

A fuel that is already in a gaseous state can assume the shape of its containment. In this respect, it is similar to the physical characteristics of a liquid. The difference is that a gas has no specific volume and is compressible. Since a gaseous fuel does not require any conversion prior to combustion, it burns more quickly than either a solid or liquid. Examples of gaseous fuels include butane and acetylene.

63. What is the underlying principle of this passage?
    A. A fuel is always recognized as being a solid, liquid or gas.
    B. The physical characteristics of fuels in their various states are investigated.
    C. How efficiently a fuel burns depends on what state it is in.
    D. The prerequisites of combustion as they relate to the various state of fuel are highlighted.

64. The temperature required to convert a fuel from a liquid state to a gaseous state is considered to be what?
    A. The pyrolysis of a fuel
    B. The flash point of a fuel
    C. Compression
    D. British Thermal Unit requirement

65. Another term for a liquid substance that demonstrates a low flash point is
    A. Butyl alcohol   B. Compressible   C. Volatile   D. Acetylene

66. How is liquid similar to a gas?
    A. Both assume the shape of their container.
    B. A liquid is essentially as combustible as a fuel existing in gaseous state.
    C. Both states can be squeezed into a smaller area.
    D. It has an inverse relationship between surface area and heat absorption.

67. On the basis of this reading, which of the substances given below would burn more readily?
    A. Coal
    B. Isopropyl alcohol
    C. Gasoline
    D. Butane

68. Which of the substances shown in the chart below poses the greatest risk of accidental combustion while being handled?

SUBSTANCE	FLASH POINT
Gasoline	-36°F
Aviation fuel	-50°F
Kerosene	100°F
Mineral oil	380°F

    A. Gasoline   B. Aviation fuel   C. Kerosene   D. Mineral oil

COMPLETE FIRE-FIGHTERS EXAM PREPARATION BOOK

QUESTIONS 69 THROUGH 77 ARE BASED ON THIS DIAGRAM

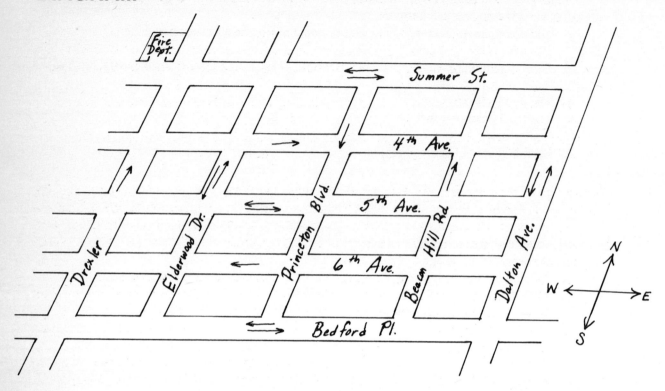

Arrows indicate direction of traffic flow

69. If the fire department needed to dispatch a medical aid unit to the corner of Drexler and 5th Avenue, what would he the most direct route to take?
    A. Left on Summer Street, right on Elderwood Drive, and then right on 5th Avenue to Drexler.
    B. Left on Summer Street and then right on Drexler to 5th Avenue.
    C. Left on Summer Street, right on Princeton Boulevard and then right on 5th Avenue to Drexler.
    D. Right on Summer Street, left on Elderwood Drive, and then left on 5th Avenue to Drexler.

70. If a fire inspector were to drive four blocks east from the fire house, three blocks south, and then two blocks west, where exactly would he or she end up?
    A. The intersection of 6th Avenue and Princeton Boulevard
    B. The intersection of 4th Avenue and Beacon Hill Road
    C. The intersection of 4th Avenue and Princeton Boulevard
    D. The intersection of Dalton Avenue and Bedford Place

71. According to the diagram, what direction is the intersection of Elderwood Drive and 6th Avenue from the intersection of Summer Street and Dalton Avenue?
    A. Northeast   B. South   C. Southwest   D. North

174

72. A fire truck was heading west on Bedford Place and had just passed Beacon Hill Road when an emergency call came over the radio. If the truck was directed to respond to a two-car collision at the intersection of Beacon Hill Road and Summer Street, what would be the most direct approach to take, considering the circumstances?
    A. Turn south on Princeton Boulevard to Summer Street and go west one block.
    B. Make a U-turn on Bedford Place and then proceed up Beacon Hill Road to the accident site.
    C. Turn north on Elderwood Drive, east on 6th Avenue, and then north on Beacon Hill Road to Summer Street.
    D. Turn north on Elderwood Drive and then east on Summer Street to the accident site.

73. Where is the fire department in relation to the intersection of Drexler and Bedford Place?
    A. South   B. North   C. West   D. East

74. Someone new to the area stops at the fire department to ask for directions. If this person is trying to get to Sheridan Village Plaza, at the southeast corner of the intersection of 5th Avenue and Beacon Hill Road, which set of directions given below would provide the most direct route? (Assume parking access is limited to an entryway located on the east side of the street between 5th and 6th Avenues).
    A. Go east on Summer Street, and then south on Beacon Hill Road to 5th Avenue.
    B. Go east on Summer Street, south on Elderwood Drive, east on 5th Avenue and then south on Beacon Hill Road half a block before making a turn east into the parking lot.
    C. Go east on Summer Street, south on Dalton Avenue, west on 6th Avenue, and then north half a block on Beacon Hill Road and then make a turn east into the parking lot.
    D. Go east on Summer Street, south on Elderwood Drive, east on Bedford Place, north on Beacon Hill Road half a block past 6th Avenue and make a turn west into the parking lot.

75. Where is Sheridan Village Plaza in relation to the southeast corner of the intersection of 4th Avenue and Princeton Boulevard?
    A. Northwest   B. Northeast   C. Southwest   D. Southeast

76. Assume for the moment that there is a residential fire on 6th Avenue between Princeton Boulevard and Beacon Hill Road. If the closest fire hydrant is located on the northwest corner of the intersection of Bedford Place and Beacon Hill Road, which set of directions given below would give the most direct approach for fire apparatus? (Take into account the necessity for quick hydrant hookup and laying of hose to the scene of the fire).
    A. East on Summer Street, south on Elderwood Drive, east on 5th Avenue, south on Dalton Avenue, and then 1½ block west on 6th Avenue.
    B. East on Sumner Street, south on Princeton Boulevard, and then half a block east on 6th Avenue.
    C. East on Summer Street, south on Princeton Boulevard, east on Bedford Place, north on Beacon Hill Road, and then half a block west on 6th Avenue.
    D. East on Summer Street, south on Dalton Avenue, and then 1½ blocks west on 6th Avenue.

77. Refer to Question 76 again. Where is the fire hydrant located in respect to the residence that is on fire?
    A. Southeast   B. South   C. Northwest   D. North

COMPLETE FIRE-FIGHTERS EXAM PREPARATION BOOK

QUESTIONS 78 THROUGH 84 ARE BASED ON THIS DIAGRAM

Numbers denote residential structures

∿∿∿ denotes fencing

78. What direction is the residence at 820 Havenor Place in relation to the fire department on Patterson?
    A. North   B. South   C. West   D. Southeast

79. By the time the fire department was notified that there is a fire at the lumber yard, it had grown to intense proportions. It was no longer a question of how to extinguish it; rather, firefighters had become more concerned with containment. Considering the fact that there was a fairly stiff wind heading south, what structure in addition to the residential units would be in jeopardy from flying ash?
    A. 1014 Mossy Park Lane
    B. Motel on Garfield
    C. 1205 Mossy Park Lane
    D. Church on Creston

176

PRACTICE EXAM 1

80. What is the number of the house located at the northeast corner of the intersection of Creston and Lagrande?
    A. 1010 Lagrande   B. 830 Creston   C. 833 Creston   D. 833 Lagrande

81. Where would someone end up after driving one block north of the police department and then going one block east before going another two blocks north?
    A. Between 830 Havenor Place and the lumber yard.
    B. In front of the hospital.
    C. In front of the motel.
    D. On Dryden just east of 810 Mossy Rock.

82. If a police unit must respond to a domestic dispute at 1213 Mossy Park Lane, what would be the most direct route for a unit located at 810 Creston to take to arrive at the scene?
    A. Proceed west on Creston and then make a left turn on to Curtis.
    B. Make a left turn on Dryden and then another left turn on Mossy Park Lane and continue approximately two and a half blocks.
    C. Make a right turn on Dryden and then a left turn on Mossy Park Lane and continue approximately two and a half blocks.
    D. Proceed south on Dryden to Mossy Park Lane and then go approximately two and a half blocks west.

83. Where is the attorney's office in relation to 833 Mossy Park Lane?
    A. Northwest   B. Southeast   C. Northeast   D. Southwest

84. Assume that the residents of 820 Havenor Place attended the church located on the corner of Lagrande and Creston. What would be the most direct route these people could take to get from their home to the church parking lot via the south entryway?
    A. West on Havenor Place, south on Lagrande for 1½ blocks, and then west into the parking lot.
    B. West on Havenor Place, south on Lagrande, and then west on Creston.
    C. West on Havenor Place, south on Curtis, and then east on Creston.
    D. West on Havenor Place, south on Lagrande, west on Patterson, north on Curtis, and then east half a block on Creston.

85. If a firefighter needed to gain entry to a premises where the doors were chained and padlocked, what tool illustrated below is most appropriate to be used to cut the chain link or the lock's shackle?

A.   B.   C.   D.

 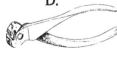

177

86. Which of the tools illustrated below would best be used to hold a piece of flat metal stock so that its end could be filed smooth?

A.   B.   C.   D.

87. The tool illustrated below is used for what purpose?

A. To etch glass
B. To pry small pieces of work apart
C. To start holes in wood to accommodate nails or screws
D. To press lines into wood with the aid of a straight edge

88. Which of the following tools below is not used in metal fabrication work?

A.   B.   C.   D.

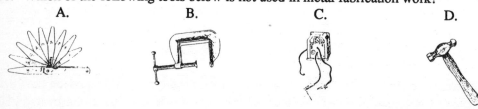

89. The tool illustrated below is principally used to cut what kind of material?

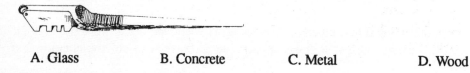

A. Glass   B. Concrete   C. Metal   D. Wood

90. Which of the following wrenches is not considered to be adjustable?
A. Crescent wrench
B. Box wrench
C. Monkey wrench
D. Stilson wrench

91. What kind of electric power tool is used to create the flared tenon joinery demonstrated in the illustration below?
A. Tenon saw
B. Reciprocating saw
C. Portable electric router
D. Portable circular saw

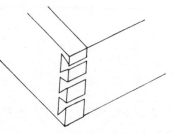

92. The device illustrated below measures what quantity?

    A. Electrical resistance
    B. Tire pressure
    C. Current flow
    D. Engine cylinder compression

93. If an electrician needed to flare the end of a piece of electrical conduit to create a coupling with another piece of electrical conduit, what kind of tool would be used?
    A. A brace with an auger bit
    B. A brace with a ream
    C. A power drill with a spade bit
    D. A power drill with a countersink bit

94. If a length of wood had to be cut perpendicular to its grain, what kind of saw blade would be recommended?
    A. Rip cut blade
    B. Saber saw blade
    C. Crosscut blade
    D. Friction blade

95. When referring to belt drives, what is the reason a belt is twisted in the manner shown to the right?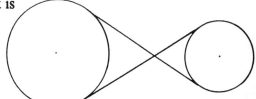

    A. It tends to increase torque on the smaller wheel.
    B. It tends to increase torque on the larger wheel.
    C. It tends to increase the speed of the smaller wheel
    D. It causes the opposing wheel to change direction.

96. Which of the gears listed below demonstrates the unique capability of changing drivetrain direction?
    A. Helical gears
    B. Beveled gears
    C. Herringbone gears
    D. Spur gears

97. According to the diagram below, at what position could a fulcrum be placed to maximize the lift capability of this lever?

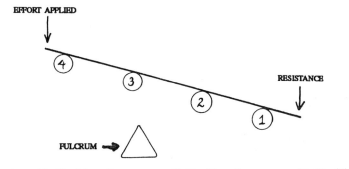

    A. Position 1   B. Position 2   C. Position 3   D. Position 4

98. Look at the diagram in question 97 again. Assuming that resistance is 825 pounds, the length of the lever is exactly 14 feet, and the fulcrum is placed dead center beneath the lever, what amount of effort would be required at the end opposite the load to attain lift?

    A. 82.5 pounds   B. 412.5 pounds   C. 825 pounds   D. 1,000 pounds

99. Judging by the solution arrived at in question 98, the use of the lever as described affords what kind of mechanical advantage for lift?

    A. 1:2   B. 2:1   C. 1:1   D. 3:2

100. According to the picture below, if 1,000 psi of pressure is exerted on the fluid trapped within the cylinder, which of the following statements is true?

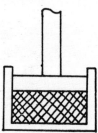

    A. That portion of fluid closest to the piston or ram comes under the greatest pressure.
    B. Pressure is evenly distributed through the fluid and against the cylinder's walls without consequent loss of power.
    C. The portion of fluid closest to the bottom of the cylinder comes under the greatest pressure.
    D. There is a proportional decrease in fluid volume.

ANSWER SHEET TO EXAM 1

PRACTICE EXAM 1

1. Ⓐ Ⓑ Ⓒ Ⓓ
2. Ⓐ Ⓑ Ⓒ Ⓓ
3. Ⓐ Ⓑ Ⓒ Ⓓ
4. Ⓐ Ⓑ Ⓒ Ⓓ
5. Ⓐ Ⓑ Ⓒ Ⓓ
6. Ⓐ Ⓑ Ⓒ Ⓓ
7. Ⓐ Ⓑ Ⓒ Ⓓ
8. Ⓐ Ⓑ Ⓒ Ⓓ
9. Ⓐ Ⓑ Ⓒ Ⓓ
10. Ⓐ Ⓑ Ⓒ Ⓓ
11. Ⓐ Ⓑ Ⓒ Ⓓ
12. Ⓐ Ⓑ Ⓒ Ⓓ
13. Ⓐ Ⓑ Ⓒ Ⓓ
14. Ⓐ Ⓑ Ⓒ Ⓓ
15. Ⓐ Ⓑ Ⓒ Ⓓ
16. Ⓐ Ⓑ Ⓒ Ⓓ
17. Ⓐ Ⓑ Ⓒ Ⓓ
18. Ⓐ Ⓑ Ⓒ Ⓓ
19. Ⓐ Ⓑ Ⓒ Ⓓ
20. Ⓐ Ⓑ Ⓒ Ⓓ
21. Ⓐ Ⓑ Ⓒ Ⓓ
22. Ⓐ Ⓑ Ⓒ Ⓓ
23. Ⓐ Ⓑ Ⓒ Ⓓ
24. Ⓐ Ⓑ Ⓒ Ⓓ
25. Ⓐ Ⓑ Ⓒ Ⓓ
26. Ⓐ Ⓑ Ⓒ Ⓓ
27. Ⓐ Ⓑ Ⓒ Ⓓ
28. Ⓐ Ⓑ Ⓒ Ⓓ
29. Ⓐ Ⓑ Ⓒ Ⓓ
30. Ⓐ Ⓑ Ⓒ Ⓓ
31. Ⓐ Ⓑ Ⓒ Ⓓ
32. Ⓐ Ⓑ Ⓒ Ⓓ
33. Ⓐ Ⓑ Ⓒ Ⓓ
34. Ⓐ Ⓑ Ⓒ Ⓓ
35. Ⓐ Ⓑ Ⓒ Ⓓ
36. Ⓐ Ⓑ Ⓒ Ⓓ
37. Ⓐ Ⓑ Ⓒ Ⓓ
38. Ⓐ Ⓑ Ⓒ Ⓓ
39. Ⓐ Ⓑ Ⓒ Ⓓ
40. Ⓐ Ⓑ Ⓒ Ⓓ
41. Ⓐ Ⓑ Ⓒ Ⓓ
42. Ⓐ Ⓑ Ⓒ Ⓓ
43. Ⓐ Ⓑ Ⓒ Ⓓ
44. Ⓐ Ⓑ Ⓒ Ⓓ
45. Ⓐ Ⓑ Ⓒ Ⓓ
46. Ⓐ Ⓑ Ⓒ Ⓓ
47. Ⓐ Ⓑ Ⓒ Ⓓ
48. Ⓐ Ⓑ Ⓒ Ⓓ
49. Ⓐ Ⓑ Ⓒ Ⓓ
50. Ⓐ Ⓑ Ⓒ Ⓓ
51. Ⓐ Ⓑ Ⓒ Ⓓ
52. Ⓐ Ⓑ Ⓒ Ⓓ
53. Ⓐ Ⓑ Ⓒ Ⓓ
54. Ⓐ Ⓑ Ⓒ Ⓓ
55. Ⓐ Ⓑ Ⓒ Ⓓ
56. Ⓐ Ⓑ Ⓒ Ⓓ
57. Ⓐ Ⓑ Ⓒ Ⓓ
58. Ⓐ Ⓑ Ⓒ Ⓓ
59. Ⓐ Ⓑ Ⓒ Ⓓ
60. Ⓐ Ⓑ Ⓒ Ⓓ
61. Ⓐ Ⓑ Ⓒ Ⓓ
62. Ⓐ Ⓑ Ⓒ Ⓓ
63. Ⓐ Ⓑ Ⓒ Ⓓ
64. Ⓐ Ⓑ Ⓒ Ⓓ
65. Ⓐ Ⓑ Ⓒ Ⓓ
66. Ⓐ Ⓑ Ⓒ Ⓓ
67. Ⓐ Ⓑ Ⓒ Ⓓ
68. Ⓐ Ⓑ Ⓒ Ⓓ
69. Ⓐ Ⓑ Ⓒ Ⓓ
70. Ⓐ Ⓑ Ⓒ Ⓓ
71. Ⓐ Ⓑ Ⓒ Ⓓ
72. Ⓐ Ⓑ Ⓒ Ⓓ
73. Ⓐ Ⓑ Ⓒ Ⓓ
74. Ⓐ Ⓑ Ⓒ Ⓓ
75. Ⓐ Ⓑ Ⓒ Ⓓ
76. Ⓐ Ⓑ Ⓒ Ⓓ
77. Ⓐ Ⓑ Ⓒ Ⓓ
78. Ⓐ Ⓑ Ⓒ Ⓓ
79. Ⓐ Ⓑ Ⓒ Ⓓ
80. Ⓐ Ⓑ Ⓒ Ⓓ
81. Ⓐ Ⓑ Ⓒ Ⓓ
82. Ⓐ Ⓑ Ⓒ Ⓓ
83. Ⓐ Ⓑ Ⓒ Ⓓ
84. Ⓐ Ⓑ Ⓒ Ⓓ
85. Ⓐ Ⓑ Ⓒ Ⓓ
86. Ⓐ Ⓑ Ⓒ Ⓓ
87. Ⓐ Ⓑ Ⓒ Ⓓ
88. Ⓐ Ⓑ Ⓒ Ⓓ
89. Ⓐ Ⓑ Ⓒ Ⓓ
90. Ⓐ Ⓑ Ⓒ Ⓓ
91. Ⓐ Ⓑ Ⓒ Ⓓ
92. Ⓐ Ⓑ Ⓒ Ⓓ
93. Ⓐ Ⓑ Ⓒ Ⓓ
94. Ⓐ Ⓑ Ⓒ Ⓓ
95. Ⓐ Ⓑ Ⓒ Ⓓ
96. Ⓐ Ⓑ Ⓒ Ⓓ
97. Ⓐ Ⓑ Ⓒ Ⓓ
98. Ⓐ Ⓑ Ⓒ Ⓓ
99. Ⓐ Ⓑ Ⓒ Ⓓ
100. Ⓐ Ⓑ Ⓒ Ⓓ

ANSWERS TO PRACTICE EXAM 1

Refer back to the reading for any clarification on Questions 1-15.

1. *C*	6. *D*	11. *B*
2. *B*	7. *D*	12. *C*
3. *A*	8. *C*	13. *D*
4. *D*	9. *A*	14. *C*
5. *B*	10. *B*	15. *D*

16. *A.* Hose streams are applied to any surface in close proximity to a fire to reduce the risk of radiant heat igniting that surface and causing a separate blaze.

17. *C.* If too much current is drawn through an outlet, the wiring to that outlet may overheat, causing the insulation to burn. If sufficient heat is given off by the wiring, a fire inside the wall may result. A short circuit in electrical wiring can also result in a fire. The Underwriters Laboratory seal of approval on any product indicates the item is safe if used for the purpose it was designed for and if the design limitations are not exceeded. However, it cannot be automatically assumed that if a product does not have UL approval that product must be unsafe. Choices B and D are false on their own merit.

18. *C.* Any kind of metal ladder poses an electrical hazard if it comes into contact with an energized power line. Aluminum is a superb conductor. Choices A and B are incorrect. Choice D refers to a problem associated with aluminum ladders. However, it is not considered a greater danger than the prospect of electrocution.

19. *B.* Problems that occur at the fire scene that hamper the efficiency of how a fire is handled are normally attributed to poor communication among firefighters. If a direction is not fully comprehended or clearly heard, it is better to have it repeated than to undertake action on your own initiative. Effective teamwork is imperative for any kind of operation to be conducted efficiently. The other choices run counter to this concept.

20. *D.* The knot demonstrated in the illustration is called a sheepshank. Its principle function is to take up slack and tighten a line. Choices A, B, and C are incorrect.

21. *D.* All of the choices given are appropriate means of handling clothing fires. Depending on the circumstances present, any one of these measures will effectively extinguish a clothing fire and prevent the victim from being burned further. Two inappropriate things that people often do when clothing catches fire are to beat at the flames or to run in panic. Either way, the flames are further fanned, making the situation worse.

22. *C.* Brakes, contrary to popular thought, are not directly responsible for stopping a vehicle. Instead, they retard the motion of the wheels. Consequently, friction develops between the tire and the road which is directly responsible for a vehicle coming to a complete stop. If ice is present, friction is reduced, which effectively impairs stopping. Glazing is a term used to describe what happens to brake linings when subjected to too much heat. Brake linings be-

come glass-like and braking capacity is minimized. Choice D is wrong because wheels do not lock up on such surfaces unless brakes are applied too hard for the road conditions. When driving on ice, it is recommended that brakes be pumped intermittently to maintain traction.

23. D. Choice A may seem to be the correct response. However, when the destructive capabilities of fire are compared to those of tornados, hurricanes, earthquakes, and floods, it is a misconception to think that fire is the worst. In fact, the destructive energy of the other phenomenon discussed is significantly greater. Additionally, fire prevention and education are taken for granted by most people, thus leading to preventable accidents and consequent fatalities.

24. B. Choices A, C, and D are procedures to follow when attempting to ventilate heat and smoke from a structure. The way choice B is worded, the vented exhaust is being directed against a prevailing wind. This is neither efficient nor desired because the vented gases could be inadvertently recirculated into the building. The best idea is to use wind to your advantage, not have it work against you.

25. A. Breaching a wall in the fashion shown in illustration A allows a wall's strength to be maintained. Choices B, C, and D will actually undermine a wall's integrity. The wall might no longer be able to bear the load it was designed to bear, thus leading to collapse and possibly injuring those working in the vicinity.

26. A. Regardless of a person's ability, if he or she does not possess a good attitude toward driving, the situation is ripe for an accident to occur. All too often, poor attitudes foster contempt and complete disregard for the rights of others. The expectation of having the right-of-way all the time can lead to a serious accident. A good attitude is an essential precursor to good driving.

27. C. It is a fact that hose lines left unused for long periods of time deteriorate more quickly than hoses that are frequently used. Unloading hose at least once a month eliminates the problem of bends becoming permanently set in the lining. Once bends happen, hoses can become weak at the bend points and are more prone to leakage or rupture. Choice A is correct if a hose length is unused for long periods of time, however, 30-day periods are too short for this effect to occur. Choices B and D are false as well.

28. D. Choices A and C are procedures that prevent a hose from becoming chafed. Among the most common causes for hoses to become damaged is allowing vehicles to cross a charged line. If it is absolutely necessary to run fire equipment over laid hose, the use of hose ramps or bridges is recommended. Choice D is a practice that can actually diminish hose life. Most hoses are composed of rubber, natural or synthetic. When rubber is subjected to hot air, it quickly deteriorates or cracks. Moderately warm air should be used to dry hose line.

29. C. A good rule of thumb for determining ladder placement is to place the heel of the ladder one-fourth the total length of the ladder from the base of what is being ascended. For example, if you had a 40-foot ladder, you would place the heel approximately 10 feet from the base of the building, tree, or whatever is being ascended. Choice B would be prone to tipping over. Choices A and D are incorrect because a ladder is not designed to bear heavy loads in this fashion. Structural failure can occur if a ladder is overextended.

30. B. Braided nylon would, of course, be the strongest rope shown; however, as it is worded, the question eliminates this as a possibility. Even unbraided 1½ inch diameter nylon can support 5,610 pounds. Polypropylene is the second strongest choice given and manila would be considered the weakest.

31. C. The weakest ⅞-inch-diameter rope shown is dacron, which is rated at 1,496 pounds. Manila would be the second weakest ⅞-inch-diameter rope having a rating of 1,540 pounds.

ANSWERS TO PRACTICE EXAM 1

32. **C.** 1½-inch-diameter dacron is 3,303 pounds stronger than ½-inch-diameter dacron (3795 - 492 = 3,303).

$$\frac{3303}{492} = 6.71, \text{ so strength was increased } 671\% \, (6.71 \times 100)$$

1½-inch-diameter Manila is 3,170 pounds stronger than ½-inch-diameter Manila (3,700 − 530 = 3,170).

$$\frac{3170}{530} = 5.98, \text{ so strength was increased } 598\% \, (5.98 \times 100)$$

1½-inch-diameter nylon is 4,884 pounds stronger than ½-inch-diameter nylon (5,610 − 726 = 4,884).

$$\frac{4884}{726} = 6.73, \text{ so strength was increased } 673\% \, (6.73 \times 100)$$

1½-inch-diameter polypropylene is 4,794 pounds stronger than ½-inch-diameter polypropylene (5,525 − 73 = 4,794).

$$\frac{4794}{731} = 6.56, \text{ so strength was increased } 656\% \, (6.56 \times 100)$$

Therefore, according to these calculations, nylon demonstrated the largest proportionate increase of strength (673%) when comparing ½-inch diameter to 1½ inch diameter. Dacron demonstrated the second-highest proportional increase of strength (671%).

33. **B.** The lowest common denominator (or LCD) for the four fractions is 16.
⁷⁄₁₆, ⅝ = ¹⁰⁄₁₆, ½ = ⁸⁄₁₆, ¾ = ¹²⁄₁₆.
Therefore,

$$\frac{7}{16} + \frac{10}{16} + \frac{8}{16} + \frac{12}{16} = \frac{37}{16} \text{ or } 2\tfrac{5}{16}.$$

34. **D.** To determine the number of bricks required, simply divide

23⅝ into 283½.

283½ ÷ 23⅝ = X

$$\frac{567}{2} \div \frac{189}{8} = X$$

$$\frac{567}{2} \times \frac{8}{189} = X = 12$$

35. **B.** F° = (⁹⁄₅)(25°) + 32 = 77°.
Always do multiplication or division first before doing any addition or subtraction, unless

there is any work in parenthesis. In that case, work the part in the parenthesis first, then do multiplication and division second, followed by any addition or subtraction. Otherwise, you will end up with choice A, which is incorrect.

$$\frac{9}{5} \times \frac{25}{1} = \frac{225}{5} = 45.$$

$$45 + 32 = 77° \text{ F.}$$

36. **D.** Under normal working conditions we can expect

$$175 \text{ hp} \times \frac{7}{8} = \frac{1225}{8} = 153\tfrac{1}{8} \text{ hp}$$

$$224 \text{ hp} \times \frac{6}{7} = \frac{1344}{7} = 192 \text{ hp}$$

This combined horsepower = $153\tfrac{1}{8} + 192 = 345\tfrac{1}{8}$ hp.

37. **A.** If an 850-liter tank was filled six times during the year, 850 liters x 6 = the number of liters consumed annually, which in this case is 5,100 liters. However, the question asks how many kiloliters are consumed annually, not liters. Since a kiloliter is 1,000 liters, we can set up the following equation to determine kiloliters.

$$5100 \text{ liters} \times \frac{1 \text{ kiloliter}}{1000 \text{ liters}} = \frac{5100 \text{ kiloliters}}{1000} = 5.1.$$

38. **C.** Since we are dealing with two different units of measure (i.e., yards and inches), it is necessary to convert one to the other. This can be done one of two ways. We know there are 36 inches in a yard, so 16 inches is equal to $^{16}\!/_{36}$ yard or, in reduced form, $^{4}\!/_{9}$ yards. Or we can change yards into inches; we simply multiply

$$1\tfrac{1}{3} \text{ or } \frac{4}{3} \times \frac{36 \text{ inches}}{1 \text{ yard}} = \frac{144}{3} = 48 \text{ inches}$$

Let's figure the ratio on the basis of inches.
$1\tfrac{1}{3}$ yards : 16 inches = 48 inches : 16 inches, or 3:1 when reduced.

39. **C.** By setting up the following proportion, we can solve for X.

$$\frac{49}{X} = \frac{42}{100}$$

$$42 X = 4,900$$

$$X = 116.67$$

40. **C.** Since we know 245 responses represents 92% of the total calls (i.e., 100% - 8% = 92%), we can set up the following proportion to determine how many calls were received altogether:

$$\frac{92}{100} = \frac{245}{X} \quad 92 X = 24{,}500;\ X = 266 \text{ calls}$$

41. **A.** This is a problem that necessitates an inverse proportion. If we try to solve this with a direct proportion, we would set it up in the following manner:

$$\frac{14.75 \text{ inch circumference}}{23.25 \text{ inch circumference}} = \frac{X \text{ rpm}}{200 \text{ rpm}}$$

$$23.25\, X = 2950;\ X = 126.88 \text{ rpm}$$

However, we know this cannot be the right answer since pulley A should have a higher rpm than pulley B because of its smaller size. So it is necessary to invert one of ratios in this proportion to find the correct answer.

$$\frac{23.25}{14.75} = \frac{X \text{ rpm}}{200 \text{ rpm}} \quad 14.75\, X = 4{,}650,\ X = 315.25 \text{ rpm}$$

42. **B.** This problem can be solved using a direct proportion.

$$\frac{100 \text{ feet of } 1\tfrac{1}{2} \text{ hose}}{150 \text{ feet of } 1\tfrac{1}{2} \text{ hose}} = \frac{25 \text{ psi}}{X \text{ psi}}$$

$$100\, X = 3{,}750\,;\ X = 37.5 \text{ psi}$$

43. **B.** This kind of problem should be set up as a direct proportion.

$$\frac{6 \text{ feet tall}}{X \text{ feet tall}} = \frac{7.5 \text{ foot shadow}}{35 \text{ foot shadow}}$$

$$7.5\, X = 210;$$

$$X \quad \frac{210}{7.5} = 28 \text{ feet.}$$

44. **D.** In this example, we already know the lengths of belts between centers (i.e., 35 cm + 35 cm = 70 cm)

Now, we must account for the pulleys' circumferences to determine the length of belt required for the ends. Since each end represents ½ the circumference, two ends would represent the whole circumference of either pulley since they are equal in diameter.

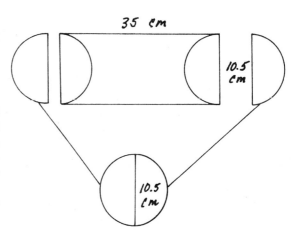

Circumference of a circle = π x Diameter
X = 3.1416 x 10.5 cm
X = 32.987 cm

Therefore, the length of V-belt required for these pulleys is equal to 35 cm + 35 cm + 32.987 cm = 102.987 cm.

45. A. What is essentially being described in this question is the volume of a cylinder. The bore represents the diameter of the cylinder and the stroke represents the height of the cylinder. Since volume of a cylinder = π x R^2 x Height, we can easily figure the displacement. The radius is equal to Diameter x .5 or, in this case, 3.25 x .5 = 1.625. (Note that 1.625^2 = 2.641) Therefore, (3.1416) x (2.641) x (4.75 inches) = X cubic inches of displacement. X = 39.41 cubic inches.

46. A. The area of a square is equal to its side squared.

    $(24 \text{ inches})^2$ = 576 square inches

The area of a circle is equal to π x R^2

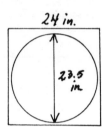

R = .5 x diameter or 23.5 x .5 = 11.75 inches
3.1416 x $(11.75)^2$ = A
3.1416 x 138.0625 = 433.74 square inches

Therefore, the area of the square minus the area of the circular portion removed will give us how many square inches of sheet metal would be left over.

576 − 433.74 = 142.26 square inches.

47. C. When determining the average of test scores for an applicant, it is necessary to add all test scores together and divide by the number of tests taken. Therefore, if we know what the four previous test scores were, and the desired overall average, we can solve the percent required in the last exam.

$$\frac{80\% + 82\% + 90\% + 87\% + X\%}{5} = 87\%$$

$$\frac{3.39 + X\%}{5} = .87$$

3.39 + X% = 4.35
X% = .96   X = 96%

48. D. This is a question that requires a compound proportion. The ratios involved are:

$$\frac{6 \text{ firemen}}{5 \text{ firemen}} , \frac{25 \text{ hydrant inspections}}{X \text{ hydrant inspections}} \text{ and } \frac{2 \text{ days}}{4 \text{ days}}$$

This can be thought of as a direct proportion because obviously the more firemen involved, the larger the number of hydrants that can be inspected. In other words, there is a direct instead of an inverse relationship. The same kind of direct relationship exists between the time and the inspections conducted (i.e., the more time involved, the greater the number of hydrants that can be inspected). Therefore, the ratio can be multiplied as shown below to set up the compound proportion.

$$\frac{6}{5} \times \frac{2}{4} = \frac{25}{X} \ ; \ \frac{12}{20} = \frac{25}{X}$$

$$\frac{12}{20} \text{ is equal to } \frac{3}{5} \text{ when reduced.}$$

$$\frac{3}{5} = \frac{25}{X} \ ; \ 3X = 125$$

$X = 41.67$ or, when rounded off, 42 hydrants could be inspected.

49. **C.** Choice C best says what the article was trying to convey. Choice A is incorrect because there wasn't any discussion within the passage regarding fire extinguisher operations. Choice B is correct for what it says; however, it does not summarize the intent of the reading. Choice D not only falls short of describing what the passage said; it also is incorrect. True, each fire is unique, but it cannot be singled out as belonging strictly to one of the four classes discussed. As mentioned in the passage, quite often a fire belongs to more than one class depending on the fuels involved.

50. **A.** Choice A is the correct answer. If you need further review, reread the passage.

51. **B or D** Class A and C fire extinguishers are not recommended for this kind of situation. If you thought Choice B was correct, you are right if your line of thinking was with regard to the oil lubricant used by drillpress operators when drilling metal stock. If it was the lubricant that caught fire and the metal stock itself was inert, Choice B could be considered possible. On the other hand, if the metal waste on the floor was the reducing agent (i.e., fuel) involved, then a Class D fire extinguisher would best be used. Choice C would be correct if the *motor* of the drill press were in danger of igniting.

52. **C.** Of prime concern here is the fact that the table saw is an electrical appliance. Unless it is de-energized, the potential for receiving a shock exists. Therefore, an extinguishing agent that has nonconductive properties, such as potassium bicarbonate, would be recommended.

53. **D.** Sodium bicarbonate is basically a nonconductive dry chemical for use on electrical fires. Choices A and B are different from sodium bicarbonate because they are wetting agents capable of electrical conductivity. Liquified compressed gas such as Halon 1211 can be used on electrical fires as well, but is not considered to be a dry chemical.

54. **A.** Class A fires. Even though the relationships of extinguishing agent and suppression effectiveness was not discussed, we can solve this question by eliminating choices. Choices B and C are wrong because the letter C indicates the capacity to be used on electrical fires. Obviously, water used to suppress a fire of this nature poses an extreme hazard of electrical shock because water is a conduct. Choice D is wrong because it was mentioned in the passage that Class D fires require special extinguishing agents. Thus, through a process of elimination, Choice A is the correct answer.

55. C. The NFPA System For Identifying Hazardous Materials. Choice A is wrong because the Emergency Response Guide adopted by the DOT was never even mentioned in the article. Choice B is incorrect because it was mentioned at the beginning of the passage that the 704 labeling system was meant for fixed facilities such as buildings or storage tanks, not materials transported by motor vehicle or rail carrier. Choice D is correct with respect to making a fire inspector's job easier to discern hazardous materials. However, it does not represent the focal issue of the article as well as Choice C.

56. D. The National Fire Protection Association 704 Labeling System.

57. B.

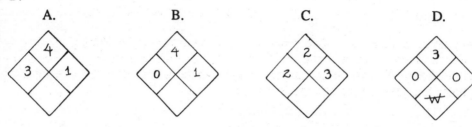

Choice A indicates a material that is extremely hazardous to health, extremely volatile or flammable, but for the most part, is fairly unreactive or stable.
Choice C indicates a material that is moderately hazardous to health, requires a moderate amount of heat for ignition and has the propensity for explosion if a heat source is involved. Choice D indicates a material that can be easily ignited because of a low flash point, in addition to the fact that it is water reactive. This placard indicates that the health hazard and reactivity of this substance is minimal.

58. A. Only Choice A comes closest to what is described as being stored in the tank. Even though there was not much elaboration on the number system, we were told that the higher the signal (i.e., number), the more hazardous the material identified. On that premise alone, Choice B can be eliminated because it would be assumed that a signal of 1 or 2 would indicate only moderate hazard posed by the contained material. Choice C is wrong because the red quadrant of a 704 label identifies flammability, not reactivity. Choice D is incorrect because the radioactive nature of the material involved is identified in the bottom quadrant of the 704 label. That does not necessarily mean the material is extremely hazardous to a person's health, extremely flammable and extremely reactive.

59. D. Yellow.

60. D. This was not mentioned in the passage.

61. B. The hazardous material spilled is reactive with water. Choice A is probably correct as far as what workers within the warehouse should do in the event of a chemical spill. However, that is not the issue indicated by the 704 label bearing a W with a slash through it. Choice C is incorrect as well because here again that is not what the label means. Choice D is wrong too because the presence of radioactivity is made known by the insignia as shown to the right in the bottom quadrant of a 704 label rather than the insignia W.

62. C. Protracting the amount of time necessary to conduct a thorough cleanup is definitely not a benefit gained by the use of the 704 label. In fact, a 704 label would expedite cleanup because knowing what the nature of the chemical involved is makes containment and cleanup safer

and more effective. Reread the last paragraph of the article to verify that Choices A, B, and D are benefits mentioned.

63. **D.** The requirements of combustion is the underlying principle of the passage given. Choices A and B are points that were discussed; however, neither provide complete insight into the article. The first sentence of the passage indicates the true focus of what the article is about. Choice C is actually a true statement, too, but the relationship of how efficiently a fuel burns with regard to the state of matter it is in was not expanded upon. There was only a brief mention of it toward the end of the passage.

64. **B.** The flash point of a fuel is the temperature required of a liquid to convert to a gaseous state or, in other words, vaporize. Pyrolysis concerns the conversion of solids to a gaseous state.

65. **C.** Any liquid that has a low flash point is considered to be volatile. Butyl alcohol is, in fact, one such substance. However, we are concerned with finding another term rather than an example. Choice B is wrong because it was stated in the article that liquids are not compressible. Choice D is an example of a gas, not a term.

66. **A.** A liquid and a gas both have the characteristic of assuming the shape of their containment. At first glance, Choice B seems correct, but a liquid must first be converted to the gaseous state prior to combustion. This conversion factor highlights the difference between a liquid and a gas when considering combustion. Choice C is false because it was established in the reading that liquids are not compressible. Choice D is not true either.

67. **D.** Butane would burn more readily than the other alternatives given, primarily because it is already in a gaseous state. The other elements shown require pyrolysis or vaporization to occur prior to combustion.

68. **B.** According to what is given in the question, aviation fuel has the lowest flash point. In fact, both aviation fuel and ordinary gasoline can exist in a gaseous state at room temperature. If vapors from these fuels flow across an ignition source, the result can be an explosion. Kerosene and mineral oil are much more stable due to their higher flash points.

69. **A.** Left on Summer Street, right on Elderwood Drive, and then right on 5th Avenue to Drexler is the most direct route. Choice B is wrong because Drexler is a one-way street. Choice C is wrong because that route takes you one block out of the way in getting to the emergency scene. Choice D is incorrect because the directions given are in error. If an aid unit made a right turn on Summer Street, it would essentially leave the map. The other directions given in Choice D are in error as well.

70. **A.** The intersection of 6th Avenue and Princeton Boulevard.

71. **C.** Southwest. Choice A would have been correct if the question had used the word *to* instead of *from*.

72. **D.** Turn north on Elderwood Drive and then east on Summer Street to the accident site. Choice A is incorrect because the fire apparatus cannot go south on Princeton Avenue. Choice B is incorrect because a U-turn is illegal. Choice C is incorrect because this route runs counter to the one-way direction on 6th Avenue.

73. **B.** North.

COMPLETE FIRE-FIGHTERS EXAM PREPARATION BOOK

74. *C.* Go east on Summer Street, south on Dalton Avenue, west on 6th Avenue, and then half a block north on Beacon Hill Road before making a turn east into the parking lot. Choice A is incorrect because going south on Beacon Hill Road runs counter to traffic flow. Choice B is incorrect for the same reason as Choice A. Choice D is wrong because you must turn east, not west off Beacon Hill Road to find parking.

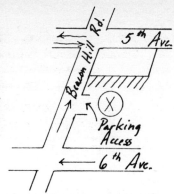

75. *D.* Southeast.

76. *C.* Going east on Summer, south on Princeton Boulevard, east on Bedford Place, north on Beacon Hill Road, and then half a block west on 6th Avenue would allow for convenient hydrant hook-up and hose lay to the scene of the fire. Choices A and D are both incorrect because the route taken on 6th Avenue is not convenient for hydrant hook-up considering the hydrant's location with respect to the fire. Choice B is incorrect because you cannot travel east on 6th Avenue.

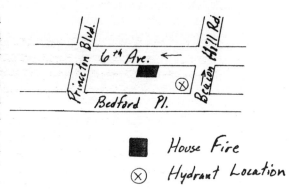

77. *A.* Southeast.

78. *A.* North.

79. *D.* Church on Creston.

80. *C.* 833 Creston. This question deserves a little more scrutiny than might be first expected. Choices A and D are incorrect because neither address exists. Look at how the street numbers get larger the further west you proceed. This pattern should indicate that these numbers apply only to streets running east–west, not north–south. Therefore, Dryden, Lagrande, Curtis, and Garfield streets do not have residential numbers, whereas Havenor Place, Mossy Park Lane, Creston, and Patterson streets do. Choice B is incorrect because that house number is located on the southeast corner of the intersection of Creston and Lagrande.

81. *B.* In front of the hospital.

82. *B.* Make a left turn on Dryden and then another left turn on Mossy Park Lane and proceed approximately 2½ blocks. Choice A is wrong because Creston is a one-way street going east. Choice C is incorrect because if a police unit were to make a right turn on Dryden from Creston, it would not intersect Mossy Park Lane. Instead, it would be heading toward Patterson. Choice D is wrong because you cannot go south on Dreyden from Creston and expect to intersect Mossy Park Lane. This option is the same route as mentioned in Choice C; however, it is worded differently.

83. *D.* Southwest.

84. *D.* West on Havenor Place, south on Lagrande, west on Patterson, north on Curtis, and then half a block east on Creston. Choice A is wrong because the question was specific about entering

192

the church parking lot via the south entryway, not the east entryway facing Lagrande. Choice B is incorrect because you cannot proceed west on Creston. Choice C is incorrect as well, because you cannot proceed south on Curtis. Both Curtis and Creston streets are one-way.

85. *B.* Bolt cutters would be the most appropriate tool used to cut chain link. Straight snips are better used to cut thin sheet metal instead of thick round metal stock. A hacksaw could cut either the lock or the chain; however, it would be significantly slower and more inconvenient to use than bolt cutters.

86. *D.* A machinist's vise would be better suited for this purpose. Channel lock pliers are handheld and can only exert as much pressure on metal stock as an individual's strength will allow. Filing the metal stock with one hand while holding it with the other is neither efficient nor safe. Pipe clamps are better suited for holding boards together for glue bonding. Parallel clamps could conceivably work by holding the metal work piece to a workbench or something of that order. However, a machinist's vise provides a better grip, thus preventing the metal workpiece from inadvertently shifting.

87. *C.* An awl is principally used to start holes in wood to accommodate screws or nails.

88. *A.* Choices B, C, and D all have various applications in the metalworking trade. A feeler gauge however, is used in the automotive trade. It is specifically used to measure the gap between various items within an engine (e.g., shaft and bearings, spark plug electrodes, etc.).

89. *A.* The tool pictured is a glass cutter.

90. *B.* The box wrench is the only alternative listed that lacks adjustability.

91. *C.* The joinery shown is a dovetail cut, created by a portable electric router. A tenon saw is a handsaw specifically designed to make similar cuts; however, the question asked for a power tool.

92. *D.* The device illustrated is a compression gauge, which quantifies engine cylinder compression.

93. *B.* This is the principle use for a ream. Choice D would seem to be a possible alternative considering how a countersink bit works on wood; however, it would be inappropriate to use it to flare electrical conduit.

94. *C.* Since the intended use of this saw blade is to cut across the grain of a piece of wood, a crosscut blade would be the better choice. Rip cut blades, on the other hand, are better suited to cut with the grain of wood. A saber saw blade is a form of blade that will fit a saber saw. It does not denote any specific type of blade per se. In other words, a saber saw blade could be a rip cut, cross cut, metal cutting, or other kind of blade.

95. *D.* Regardless of the wheel configuration involved, anytime a belt between two wheels is crossed in the manner described, it causes a directional change of the opposing wheel. Instead of both wheels turning clockwise, one wheel will turn counterclockwise. Torque and speed are not effected.

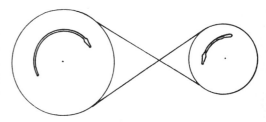

96. B. Beveled gears are unlike Choices A, C, and D in that teeth are cut into the edging of the gear, rather than set perpendicularly to a gear's facing. This angulation allows for the directional change of a drivetrain.

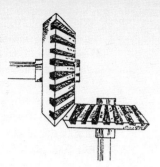

97. A. A fulcrum placed at Position 1 would enhance the lift capability of this lever the most. Essentially, the closer a fulcrum is placed to a load or resistance, the easier it becomes to lift with the lever.

98. C.
$$\text{Effort} = \frac{825 \text{ lbs.} \times 7 \text{ feet}}{7 \text{ feet } (14 \text{ feet } - 7 \text{ feet})}$$

Effort x Effort Distance = Resistance x Resistance Distance

$$\text{or Effort} = \frac{\text{Resistance} \times \text{Resistance Distance}}{\text{Effort Distance}}$$

Therefore, Effort = $\frac{825 \text{ lbs.} \times 7 \text{ feet}}{7 \text{ feet}}$ = $\frac{5775 \text{ lbs./feet}}{7 \text{ feet}}$ or 825 lbs.

Dead center of a 14 foot lever places the fulcrum 7 feet from the load resistance.

99. C. Since it requires 825 pounds of effort to lift 825 pounds, no mechanical advantage is gained by using the lever as described. Ratios of 1:1 do not demonstrate mechanical advantage.

100. B. When a fluid trapped in a confined space is submitted to pressure, that pressure is distributed evenly in all directions without sacrifice to power. In other words, liquids are not compressible as are gases. The remaining choices are false.

TEST RATINGS ARE AS FOLLOWS:

95–100 correct	EXCELLENT
87–94 correct	VERY GOOD
81–86 correct	GOOD
75–80 correct	FAIR
74 or below correct	UNSATISFACTORY

Go back to the questions you missed and determine if the question was just misinterpreted for one reason or another, or if it reflects a particular weakness in subject matter. If it is a matter of misinterpretation, try reading the question more slowly while paying close attention to key words such as *not, least, except, without,* etc. If, on the other hand, you determine a weakness in a particular area, don't despair. That is what this book is for; to identify any type of weakness before you take the actual exam. Reread the material on the area of concern in this book and if you still feel a need for supplemental material, your local library is an excellent source.

PRACTICE EXAM 2

The time allowed for the entire examination is 2 hours.

DIRECTIONS: Each question has four choices for answers, lettered A, B, C, and D. Choose the best answer and then, on the answer sheet provided, find the corresponding question number and darken with a soft pencil the circle corresponding to the answer you have selected.

STUDY THE FOLLOWING SKETCH FOR ONLY 5 MINUTES. When your time is up, turn to the next page to answer questions 1 through 15 relating to the sketch you have studied. YOU MAY NOT LOOK BACK AT THE SKETCH. NOTES ARE PROHIBITED.

REMEMBER: This is a memory exercise, so any review of this sketch is prohibited on the actual examination. Thus, any review on the practice exam will skew your test results.

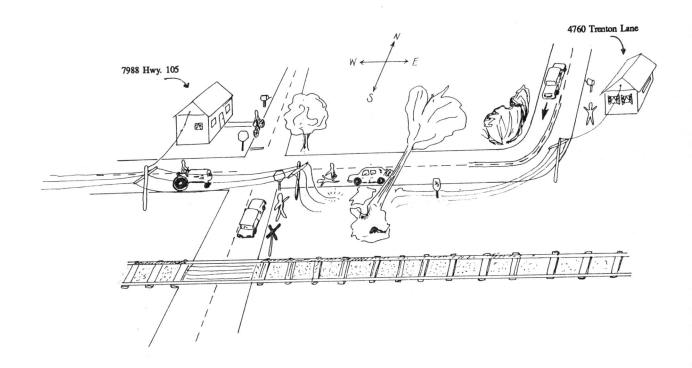

COMPLETE FIRE-FIGHTERS EXAM PREPARATION BOOK

ANSWER QUESTIONS 1 THROUGH 15 ON THE BASIS OF THE SKETCH

1. The vehicle involved in the accident was headed in what direction?
    A. North B. East C. Southeast D. West

2. What is the street address of the residence nearest the accident?
    A. 7899 Highway 105
    B. 4706 Trenton Lane
    C. 4070 Trenton Lane
    D. 7988 Highway 105

3. How many people appear injured in the sketch?
    A. 1
    B. 2
    C. 3
    D. There were no apparent injuries.

4. How many people appear to be assisting the accident victim(s)?
    A. No one was depicted in the immediate area.
    B. 1
    C. 2
    D. 3

5. Where is the railroad crossing in relation to the accident?
    A. Southeast B. South C. Southwest D. West

6. How many road signs were seen in the sketch?
    A. 4 B. 1 C. 2 D. 3

7. What other conditions exist that render the accident scene more hazardous?
    A. Any leaking fuel or oil could be ignited.
    B. Potential electrocution from a downed power line.
    C. An oncoming vehicle rounding a blind curve on Trenton Lane.
    D. All of the above.

8. The residence that experienced a power outage as a result of the tree falling across the wires is:
    A. 4670 Trenton Ln
    B. 4760 Trenton Ln
    C. 7989 Highway 105
    D. 7988 Highway 105

9. At the intersection of Highway 105 and Trenton Lane, who has the right of way?
    A. Bicyclist
    B. Tractor
    C. Car headed north on Highway 105
    D. Pedestrian in the crosswalk

10. Trenton Lane as shown by the sketch is a
    A. Two-lane, one-way road running west
    B. Two-lane, one-way road running east
    C. Two-way street
    D. Two-lane road with good shoulders

11. What is the speed limit on Trenton Lane?
    A. 35 mph   B. 30 mph   C. 25 mph   D. 40 mph

12. How many windows can be seen on the home fronting the highway?
    A. 2   B. 5   C. 4   D. 3

13. How many power line poles are in the sketch?
    A. 2   B. 3   C. 4   D. 5

14. What is the posted speed limit on Highway 105?
    A. 25 mph
    B. 45 mph
    C. It wasn't posted.
    D. The sign was partially obscured by a tree.

15. Where in the sketch is a no passing zone?
    A. The Trenton Lane curve
    B. The north side of the railroad tracks on Highway 105
    C. The south side of the railroad tracks on Highway 105
    D. That portion of Trenton Lane that intersects Highway 105

16. What is the principle reason firefighters open and close all valves and nozzles slowly?
    A. It is important to approach any emergency scene in a slow and methodical manner.
    B. It is important to appear to be in control of the situation as far as bystanders are concerned.
    C. It conserves water.
    D. It circumvents potential damage to equipment.

17. Frequently, firefighters must break windows to gain entry into a building. Keeping safety in mind, which of the following procedures would not be recommended?
    A. Stand to the leeward side of the window being broken.
    B. Keep your hands above the point of impact.
    C. Never use your hands to break glass.
    D. Initially break the topmost portion of a window, and then work your way to the bottom of the pane.

18. If a firefighter en route to an emergency became involved in a traffic accident, which of the options given below would be the next appropriate step to take?
    A. Stop immediately.
    B. Assess the amount of damage done. If it seems insignificant, it would simply be a waste of time to report it.
    C. Notify his or her superior after the police have conducted a thorough investigation.
    D. Bring the vehicle to a halt only when it is safe to do so.

19. When artificial respiration is required, the victim's head is tilted back by lifting gently on the back of the neck and pushing the forehead down. What is the principle reason for doing this?
    A. It opens the throat for free air passage.
    B. It prevents the victim from making any unnecessary movement.
    C. It lessens the degree of shock the victim will suffer.
    D. It allows for better blood circulation to the victim's brain.

20. The protective barrier between a car's engine and the passenger compartment is referred to as a fire wall. Judging by its location, this barrier principally serves what function?
    A. It protects the engine from a passenger compartment fire.
    B. It guarantees the safety of passengers if the engine catches fire.
    C. It will impede the spread of flames from the engine to the passenger compartment.
    D. It prevents the gas tank from exploding.

21. Gasoline engines become more efficient when they operate at sea level versus higher elevations. What is the most probable reason for this?
    A. There is less frictional loss at sea level.
    B. The air is thinner at higher elevations.
    C. The air is thicker at higher elevations.
    D. The air is cooler at sea level.

22. The illustrations below represent four different hose lay configurations. Each coupling shown represents a connection between 40-foot lengths of 2½-inch hose. Which hose lay would experience the most pressure loss, assuming flow pressures from the hydrants were the same for each?

    A.   B.   C.   D.

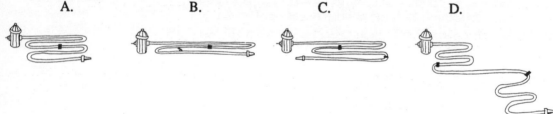

23. Referring to the same illustration seen in question 22, which selection would yield the greatest flow pressure at the nozzle?
    A. A   B. B   C. C   D. D

24. On occasion, fire departments receive emergency calls regarding people that have fallen through the ice on frozen lakes or ponds. Throwing a rope or line to the victim is sometimes ineffective because the victim panics or because hypothermia precludes any possibility of a safe rescue. When firefighters lay a wooden ladder on the ice and use it as a crawlway to extend direct reach to the victim, what would this be considered?
    A. Safe; even if the ice were to break, the ladder would float.
    B. Unsafe. It would be quicker to just walk to the edge of the ice and pull the victim out of the water.
    C. The safest alternative considering the circumstances. The ladder serves to displace weight over a wider area and thus makes it less likely that the ice will break further.
    D. Unsafe. The added weight of the ladder may cause the ice to break further, thus complicating effective rescue.

25. Without cutting the tarp, what is the largest square area that could be covered by a salvage tarp measuring 10 feet x 15 feet?
    A. 100 square feet
    B. 150 square feet
    C. 200 square feet
    D. 250 square feet

26. Which of the firefighting practices below give the greatest benefit in reducing the need for extra manpower when salvage operations are undertaken?
    A. Open all doors and windows to provide extensive ventilation.
    B. Use a hose stream judiciously when suppressing a fire.
    C. Remove as much furniture or merchandise within the building as quickly as possible.
    D. Cover everything possible with polyethylene sheeting to reduce further damage from smoke.

27. Which kind of fire is probably the worst for firefighters to cope with?
    A. Rural wildlands fire.
    B. Enclosed mall or shopping center fires.
    C. Hospital or nursing home fires
    D. Suburban residential fires.

28. What is the reason that air should be bled from a hose immediately before advancing into a burning structure?
    A. The hose may rupture or blow up.
    B. The rush of air out of the hose could stir up ash and restrict visibility.
    C. It lessens the possibility of the hose line getting bound in a door jam.
    D. It insures that water will be available immediately upon demand.

29. Electing to use a fog or fine spray instead of a solid stream does not provide which of the following advantages?
    A. This method uses less water.
    B. It covers a greater area.
    C. It absorbs more heat.
    D. It requires low discharge pressure.

30. Upon arriving at a warehouse suspected of having a gas leak of unknown origin inside, two things were immediately apparent to the fire crew. Both metal doors to the front and back were locked and there were no windows on the ground floor. Under the circumstances, what is the best possible means to conduct a forcible entry?
    A. Use an oxyacetylene cutting torch on the front door.
    B. Utilize a pry bar or some similar tool to wedge open the door at the jamb.
    C. Use a fireman's axe to create a hole in the door large enough to accommodate a hand and unlock the door from the inside.
    D. Utilize a gasoline powered circular saw with a metal cutting blade.

31. If a driveshaft made 40⅝ revolutions every 2½ minutes, what would be the shafts rpm (revolutions per minute)?
    A. 13⅓ rpm   B. 15³⁄₇ rpm   C. 16¼ rpm   D. 87 rpm

32. To determine the total current of a particular circuit, individual currents rated in amperage must be added together. What is the total current of a circuit that has three currents that measure 1³⁄₁₀ amps, 2¹²⁄₁₀₀ amps and 1¹⁷⁄₁₀₀₀ amps?
    A. 4⁴³⁷⁄₁₀₀₀ amps   B. 4³²⁄₁₁₁₀ amps   C. 4¹⁷⁄₁₀₀₀ amps   D. 4 amps

33. What is the perimeter length of the trapezoid shown below?

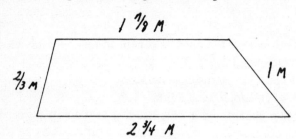

    A. 4¹²⁄₁₅ meters   B. 5³⁄₁₆ meters   C. 5¹⁄₆ meters   D. 6⁷⁄₂₄ meters

34. If a booster line (i.e., rubber covered, rubber lined hose) had an internal diameter of ¾ inch and the thickness of the hose wall is ¹⁄₁₆ of an inch, what is the booster line's outside diameter?
    A. ⁵⁄₃₆ inch   B. ¹³⁄₁₆ inch   C. ⅞ inch   D. ⅚ inch

35. How many centimeters are there in 2.67 meters?
    A. 2,670   B. 267   C. 26.7   D. 2.67

36. A firehose coupling has an outside diameter of 2⅞ inches and an interior diameter of 2⅓ inches. What is the thickness of the coupling wall?
    A. 697 inches   B. 0.65 inches   C. 0.542 inches   D. 0.271 inches

37. Ohm's Law deals with the relationship of three variables.

$$R = \frac{E}{I}$$ where R = Resistance in ohms
    I = Current in amps
    E = Pressure in volts

    How could this equation be written to find E?
    A. $E = R/E \times I$ B. $E = IR$ C. $E = E/R \times I$ D. $E = I/E \times R$

38. When speaking of an engine's thermal efficiency, the amount of heat transformed into work is compared to the amount of heat produced. This is normally measured in British Thermal Units or BTU. If an engine produces 72,500 BTU, of which 60,000 BTU transformed into work, what is the thermal efficiency of this engine expressed as a percentage?
    A. 120.83%   B. 115.7%   C. 93.28%   D. 82.75%

39. On the basis of the principle described in the previous question, how many BTU can we assume are transformed into work if we know that an engine is 61.3% thermally efficient and produces a total of 87,500 BTU?
    A. 47,960.5 BTU   B. 53,637.5 BTU   C. 57,960.5 BTU   D. 82,600 BTU

40. If the two pulleys shown have a 3:1 ratio and the smaller pulley has a 9.5 inch circumference, how fast would the larger pulley turn if the smaller pulley turned at a rate of 300 rpm?

   A. 900 rpm B. 300 rpm

   C. 100 rpm D. 75 rpm

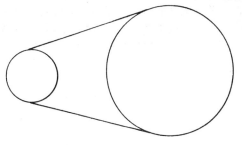

41. What is the diameter of a fire extinguisher that has a circumference of 19¼ inches?
    A. 8.62 inches   B. 7.8 inches   C. 6.127 inches   D. 5.911 inches

42. How high would a 25-foot ladder reach if it is placed 6 feet from the base of a building?
    A. 25 feet   B. 24.27 feet   C. 23.12 feet   D. 22 feet

43. How many square feet of tarp would be needed to cover a circular area that has a radius of 7.35 meters?
    A. 46.18 feet
    B. 153.1 square feet
    C. 169.72 square feet
    D. 173 square feet

44. If someone had an attic that measured 48.5 feet in length and 22.5 feet in width and was 12 feet high, how many cubic feet of space could be utilized for storage?
    A. 7,905.2 cubic feet
    B. 6,480 cubic feet
    C. 967 cubic feet
    D. 135 square feet

45. A solution contains 35% benzene and 65% water. If 14 liters of this solution is diluted by adding 4.5 liters of water, what percentage of benzene would be present in the new mixture?
    A. 23.7%   B. 26.49%   C. 31.7%   D. 43.55%

46. Which of the expressions below would represent the product $(X + 3)(X - 8)$?
    A. $X^2 - 5X - 24$ B. $2X - 5X - 24$ C. $X^2 + 3X - 8$ D. $X^3 - 5$

ANSWER QUESTIONS 47 THROUGH 52 ON THE BASIS OF THE FOLLOWING PASSAGE

Static electricity is a natural phenomenon that by itself can pose little, if any, danger. However, if it is present in an area of stored flammable substances, a real danger exists. Normally, a nonconductive material has a charge present on its surface. Some of the time this charge can diminish by itself. Other times, if there is an insufficient electrical path for this charge to flow along, an open spark can be the consequence. Therein lies the extreme danger of potential ignition of flammable materials.

Static electricity cannot be totally eliminated but steps can be taken to minimize the potential danger. Humidity can play an important role in controlling static electricity. The higher the humidity, the less chance there is for static electricity to manifest itself as an open spark. In some instances, this may not be a practical option.

Another alternative is grounding the affected surfaces with a conductor such as a wire or cable. This way a good electrical path is provided which essentially siphons off any potential charge.

One other means of dealing with static electricity is bonding. The surfaces of two or more objects are interconnected with a conductor. Unlike a ground, the static charge is not removed but the overall charge is dissipated between the connected surfaces.

A more complicated means of controlling static electricity is employing a device that ionizes the air. These devices bleed off charges on affected surfaces. This technique is used widely in various industrial capacities.

As a final form of control, an open flame is noted for diminishing static electricity potential. For obvious reasons though, this method is not utilized where flammable material is present.

47. This reading is chiefly concerned with what kind of problem?
    A. Static electricity
    B. Flammable materials
    C. The potential for electrical shock
    D. The prospect of an electrical spark between two nonconductive surfaces in the presence of flammable material.

48. Of the options discussed that can control static, which one is the most impractical in the presence of flammable liquid storage?
    A. Air humidification
    B. Air ionization
    C. Open flame
    D. Bonding

49. According to the passage, which of the following statements is true?
    A. Static electricity is a natural phenomenon that cannot be entirely eliminated.
    B. The higher the humidity, the greater the prospect of a static spark.
    C. To properly ground a surface, a nonconductive lead should be used.
    D. Bonding is a process whereby air in a statically charged environment is ionized.

50. When inspecting a chemical storage facility, John Talbot noted that a single copper wire was attached by means of a clamp to all of the ten cylindrical tanks present. This would constitute what means of static electricity control?
    A. A ground   B. Bonding   C. Ionization   D. All of the above

51. What condition needs to be present for static electricity to manifest itself as a spark?
    A. A fairly dry environment
    B. Nonconductive materials that possess different charges
    C. An insufficient electrical path
    D. All of the above

52. What do air ionizers used by some businesses essentially do?
    A. Serve to humidify a workroom environment
    B. Provide an exit for electrical charges that build up on various surfaces
    C. Reduce the potential danger of a lightning strike
    D. Minimize the possibility of an employee receiving a lethal dose of static conductivity

ANSWER QUESTIONS 53 THROUGH 58 ON THE BASIS OF THE FOLLOWING PASSAGE

One of the most important aspects of a storage vessel used to contain various flammable liquids is the venting system. Depending on the vaporization properties of a particular liquid, it is imperative that this potential excess pressure be vented. This is especially true if an above-ground tank is to be exposed to an external heat source. As the liquid heats up, an internal pressure would build up due to vaporization. If there is no means of releasing this excess pressure, eventually the tank's integrity would be compromised, resulting in a severe explosion. In firefighting terms, it is referred to as a BLEVE (Boiling Liquid Expanding Vapor Explosion).

Therefore, venting for a storage vessel is extremely important, and consequently mandated by various regulations. Where a vent is placed on a storage vessel is important as well. Any vaporized material that is vented should not potentially impinge on other nearby storage vessels. This precaution inhibits the possibility of creating a chain reaction fire if the vented material were to ignite. The vents must also be located in a place that minimizes the risk of inadvertent damage or accidental plugging. Routine inspection of these vents should be conducted to be sure they are operating properly.

53. Which of the following choices would be an appropriate title for this passage?
    A. The Physical Dynamics of a Storage Tank.
    B. The Potential Hazards of an Improperly Vented Storage Vessel.
    C. The Fundamentals of Ventilation Systems on Storage Tanks.
    D. The Complicity of Vaporization on Unvented Storage Vessels.

54. What is the technical term for a liquid storage tank rupturing due to excessive internal pressures from vaporization?
    A. BLEAVE
    B. Improper storage tank ventilation (i.e., ISTV)
    C. Tank integrity compromise
    D. Boiling liquid expanding vapor explosion

55. What process is chiefly responsible for creating internal pressure within a liquid-filled storage tank when it is subjected to heat?
    A. Vaporization
    B. Condensation
    C. Pyrolysis
    D. None of the above

56. Hypothetically, if a liquid storage tank's ventilation system ceased to function as its contents were being drawn out via a pump, what would be the most likely result?
    A. The storage vessel would probably remain unaffected.
    B. Providing the draft was strong enough, it might collapse the tank.
    C. The storage tank would probably expand to the point of rupturing.
    D. The pump would probably fail.

57. Of the diagrams shown below, which one demonstrates the better placement of a vent? Assume the tanks diagrammed are ¾ full.

58. According to the reading, how often should ventilation systems of liquid storage vessels be checked to see if they are functioning properly?
    A. Only a routine inspection was mentioned without regard to any particular time frame.
    B. Once a week
    C. Once a month
    D. Biannually

ANSWER QUESTIONS 59 THROUGH 63 ON THE BASIS OF THE FOLLOWING PASSAGE

Any industry that has a fabrication or refinement process which yields combustible dust as a by-product is faced with a special problem: Namely, if that dust comes into contact with an ignition source, a fire or explosion may result. Steps must be taken to minimize this risk. The most direct way of coping with this problem is to enclose, as much as possible, those areas where dust originates. This may require the use of chutes that contain conveyors or augers, or boxed-in enclosures for hoppers, etc. This method essentially prevents dust from dispersing. In some applications, this may be impractical and dust may be allowed to disperse, but not accumulate to any substantial degree. Regular maintenance with a vacuum cleaning device or simply brooming the dust away by hand are two methods of control. Special care should be given not to create a dustier situation than necessary while cleaning. Combustible particles suspended in air are much more susceptible to ignition than dust particles in a pile.

Another precautionary step to prevent combustion includes making sure that all electrical equipment is properly grounded. Dust particles suspended in air can create a static electrical charge. Without good grounding, a static spark may occur and provide an ignition source for the combustible dust present. Additionally, it is important that any electrical motors in areas affected by dust be equipped with bearings and other materials that do not create sparks and that minimize friction. Following these few simple rules can significantly enhance fire safety for those that work in dusty environments.

59. What was the underlying principle of this article?
    A. Dust is a necessary evil to most industry and thus can be ignored.
    B. Dust should be regularly removed if it is considered to be combustible.
    C. How to best alleviate the dangers posed by dust in a work-related environment.
    D. The susceptibility of dust to combustion when suspended in air.

60. Which of the following would be the safest way of utilizing a fan for ventilating a dusty workplace?
    A. Replace as much of the fan's working parts with material that is less prone to spark and friction.
    B. Place the fan's motor on the exterior side of the building or, if that is not possible, arrange it such that the motor is on the cleanest side of the area affected.
    C. Be certain that the fan is properly grounded.
    D. All of the above.

61. Which of the alternatives below is the most direct means of coping with a dust problem experienced by some industries?
    A. Utilize an air humidification device to minimize the potential hazard of static spark.
    B. Be sure that all equipment has sufficient grounding.
    C. Contain the dust as best as possible at its source.
    D. Strictly adhere to a no-smoking policy for all employees working in affected areas.

62. Four janitorial crews were assigned to clean up sawdust at a cabinet manufacturing facility. Which of the various procedures described below would be considered the most precarious with regard to fire safety?
    A. Use an air blowing device to remove sawdust off of various surfaces.
    B. Use water sprays to wet down the sawdust prior to removal.
    C. Whisk and floor broom the sawdust into small piles and remove it with dustpans.
    D. Use a vacuum cleaner that has good filtration.

63. What is the most probable reason that combustible particles suspended in air are more susceptible to flammability than those left in a pile?
    A. Static charge is more prevalent in air-suspended particles.
    B. Spontaneous combustion becomes a menacing factor, depending on how fine the dust involved is.
    C. Regardless of the combustible material involved, there is a chemical transformation that takes place when particulate matter is suspended in air. This newfound formation enhances the flammability of the substance involved.
    D. When particulate matter is suspended in air, a proportional increase of surface area manifested. One consequence is the increased potential of flammability.

COMPLETE FIRE-FIGHTERS EXAM PREPARATION BOOK

ANSWER QUESTIONS 64 THROUGH 76 ON THE BASIS OF THE FOLLOWING DIAGRAM

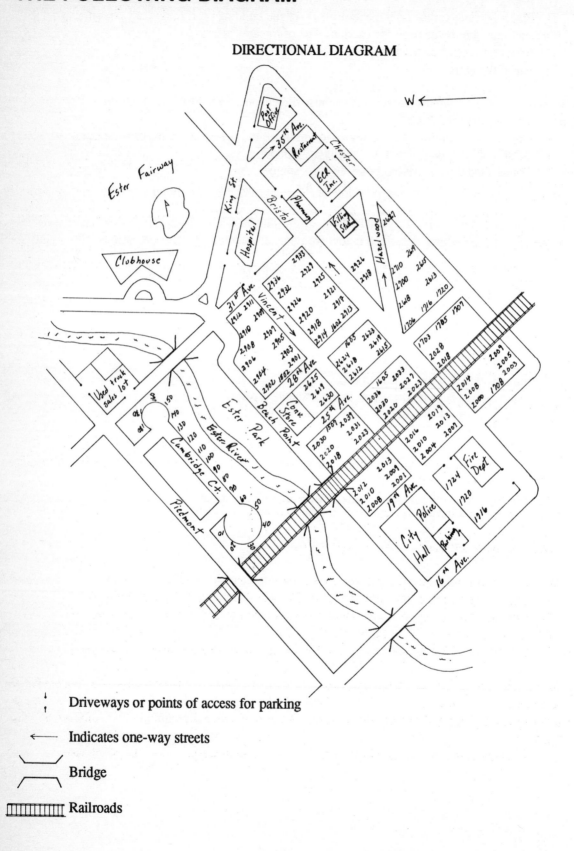

DIRECTIONAL DIAGRAM

↑ Driveways or points of access for parking

← Indicates one-way streets

⌒ Bridge

▯▯▯▯ Railroads

Facts to consider: As a general rule of thumb, city blocks that have a square or rectangular symmetry have fire hydrants located at the east corner. City blocks that have a right triangular symmetry have fire hydrants located at the corner that forms the right angle of the block. The water flow in the Ester River runs in a southeasterly direction. The intersection of Beach Point and 25th Avenue is closed between the hours of 10 a.m. to 5:30 p.m. due to the local gas utility installing an underground main. Union and Southern Pacific Railways run on the hour from 6 a.m. to 7 p.m. (i.e., 6 a.m., 7 a.m., etc.).

64. According to the directional diagram, what direction is the hospital in relation to the fire department?
    A. West   B. Southeast   C. Northwest   D. Southwest

65. If a golfer playing on the northeast side of Ester Fairway suddenly collapsed from a supposed heart attack, what would be the most direct approach to take for a medical aid unit dispatched from the fire department?
    A. Go northwest on Bristol about half a block past King Street. The victim should be within a short distance southwest of this location.
    B. Leave the fire station via the 19th Avenue exit, drive southwest to Beach Point, and then make a northwest turn on Beach Point. Stay on Beach Point Drive and go directly to the Ester Fairway Clubhouse to administer aid to the victim.
    C. Leave the fire station via the 19th Avenue exit, and then make a right turn on Vincent and continue northwest to Ester Fairway.
    D. Leave the fire station via the 19th Avenue exit, make a left turn on Chester, and go to 35th Avenue before heading southwest to the Fairway. The victim should be very close to the intersection of King Street and 35th Avenue.

66. What direction is the pharmacy in relation to the hospital?
    A. North   B. West   C. Southwest   D. Northeast

67. If the residence at 2619 Bristol had an attic fire, what fire hydrant would be the nearest for firefighters to hook up to initially?
    A. The hydrant located in front of 1706 25th Avenue.
    B. The hydrant located in front of 2615 Bristol.
    C. The hydrant located in front of 2623 Bristol.
    D. The hydrant located in front of 2033 Bristol.

68. According to the diagram, which of the following statements is false?
    A. The two streets that run parallel to Ester Park are Piedmont and Beach Point.
    B. Bristol, 35th Avenue, Vincent, and Hazelwood are the only one-way streets shown.
    C. Hazelwood runs in a northerly direction.
    D. Cambridge Court is basically north of the Ester Fairway Clubhouse.

69. At 4:30 p.m. a commuter going home from work lost control of his vehicle and collided with a telephone pole on the northeast side of Chester half a block southeast of 31st Avenue. Assuming this individual simply drove off the shoulder of the road, what direction can we surmise he was heading?
    A. Southeast   B. Northwest   C. Southwest   D. Northeast

70. What would be the most direct route for firefighters to take from the fire station in responding to the accident described in question 69?
    A. Go southwest on 19th Avenue to Chester and then head southeast on Chester to the accident scene.
    B. Go northwest on Vincent to 31st Avenue and then drive northeast to Chester.
    C. Go northeast on 19th Avenue to Chester and then go northwest to the accident scene.
    D. Go northeast to 19th Avenue to Bristol, northwest on Bristol, and then north on Hazelwood to Chester.

71. Paramedics needed to transport a middle-aged man experiencing breathing difficulties from 1708 19th Avenue to the hospital emergency room. What would be the most appropriate route to take to get to the hospital if it were one o'clock in the afternoon?
    A. Head northwest on Chester to 31st Avenue and then go southwest on 31st Avenue to the hospital entrance opposite Vincent.
    B. Go to Bristol, then go northwest on Bristol to 31st Avenue to the hospital entrance opposite Vincent.
    C. Drive southwest on 19th Avenue to Beach Point, go northwest on Beach Point to 31st Avenue, and then go northeast on 31st Avenue to the hospital entrance opposite Vincent.
    D. Start off southwest on 19th Avenue to Beach Point, southeast on Beach Point to 16th Avenue, southwest on 16th Avenue to Piedmont, northwest on Piedmont to 31st Avenue and then northeast on 31st Avenue to the hospital entrance opposite Vincent.

72. If someone who lives at 2700 Hazelwood wanted to walk to a neighbor's house located at 2027 Bristol, what would be the most direct route to take?
    A. South on Hazelwood to Bristol and then to the second house on the left southeast of 25th Avenue.
    B. North on Hazelwood and then make a sharp right turn on to Chester, walk southeast on Chester to 19th Avenue, southwest on 19th Avenue to Bristol and northwest on Bristol to 2027, and then turn left onto the drive.
    C. South on Hazelwood to Bristol and then go to the second house on the right southeast of 25th Avenue.
    D. North on Hazelwood, southwest on 31st Avenue to Vincent, southeast on Vincent to 19th Avenue, northeast on 19th Avenue to Bristol, and then northwest on Bristol to 2027.

73. One afternoon, two teenagers went into the Ester River for relief from the summer heat. It wasn't long before they realized the current was too strong for them. An alert citizen noticed the event unfolding northwest of the 31st Avenue bridge. A call was immediately put through to the police and fire departments. The response time to the scene was less than 2 minutes. At which of the following areas should rescue efforts be concentrated?
    A. Northeast of the used truck sales lot
    B. Below the 31st Avenue bridge
    C. Below the train trestle crossing the Ester River
    D. Below the 16th Avenue bridge

74. Where is the post office in relation to the ECR Inc. building?
    A. Northwest   B. Southwest   C. Southeast   D. North

75. Which residence in the block bounded by 25th Avenue, Hazelwood, and Chester streets is most distant from the fire hydrant located on that block?
    A. 2820 Hazelwood   B. 1706 25th Avenue   C. 1720 25th Avenue   D. 2627 Chester

76. If a person who lived at 120 Cambridge Court wanted to drive to the post office, what would be the shortest route he or she could take to get to the northeast entrance for parking?
    A. Go northwest on Piedmont to 31st Avenue, northeast on 31st Avenue to Bristol, northwest on Bristol to King Street, and then north on King Street to the first entrance on the right.
    B. Go northwest on Piedmont to 31st Avenue, northeast on 31st Avenue to King Street, north on King Street to 35th Avenue, northeast on 35th Avenue to Chester, and then northwest on Chester to the entrance on the left.
    C. Go southeast on Piedmont to 31st Avenue, northeast on 31st Avenue to King Street, north on King Street to 35th Avenue, northeast on 35th Avenue to Chester, and then northwest on Chester to the entrance on the left.
    D. Go southeast on Piedmont to 28th Avenue, northeast on 28th Avenue to Bristol, northwest on Bristol to 35th Avenue, northeast on 35th Avenue to Chester, and then northwest on Chester to the entrance on the left.

77. Which instrument below could best determine if the interior diameter of two pipes is identical?

    A.   B.   C.   D.

78. The instrument illustrated below is a

    A. Depth gauge   B. Spring dividers   C. Vernier calipers   D. Screw pitch gauge

79. Which of the tools below lacks the ability to miter square?
    A. Try square
    B. Combination square
    C. Carpenter's framing square
    D. Both A and C.

80. If the head of a screw had the kind of configuration illustrated below, what kind of screwdriver would be required?

    A. Phillips screwdriver
    B. Torx screwdriver
    C. Standard screwdriver
    D. Offset screwdriver

81. Which of the wrenches below is more convenient and can save time when it is necessary to loosen or tighten a bolt in tight quarters?
    A. Crescent wrench
    B. Stilson wrench
    C. Offset wrench
    D. Socket wrench

82. What kind of wrench demonstrates an L-shaped hexagonal design that is used to tighten or loosen machined setscrews?
    A. Allen wrench
    B. Torque wrench
    C. Stilson wrench
    D. Offset wrench

83. What kind of file would be better suited to sharpen the dull edges of a saw blade and achieve proper angle set?
    A. Flat file
    B. Round file
    C. Triangular file
    D. Bench grinder

84. Considering how closely buildings were spaced together, the kind of materials used in their construction, and the inadequacy of firefighting equipment in the late 1800s and early 1900s, it is not hard to understand why "conflagrations" were not that uncommon. "Conflagrations" most nearly means
    A. Firefighter strikes
    B. Out-of-control fires
    C. Fire-related medical disabilities
    D. Firefighter fatalities

85. If firefighter recruits have any degree of "acrophobia," their ability to perform their duties at a fire can suffer and put co-workers at risk. "Acrophobia" most nearly means
    A. Stagefright
    B. Fear of heights
    C. Paranoia
    D. Uncertainty

86. Some of the questions on the lieutenant's exam made Carl feel a shade of "ineptitude." "Ineptitude" most nearly means
    A. Confidence
    B. Hostility
    C. Competence
    D. Lack of aptitude

87. The structure was quite "dilapidated" and consequently it presented a host of fire hazards to the occupants. "Dilapidated" most nearly means
    A. New
    B. Deteriorated
    C. Cheap
    D. Worthless

88. If a metal workpiece is said to be "malleable," it most nearly means it is
    A. Brittle
    B. Hard
    C. Capable of being shaped
    D. Warped

89. By observing how extensive the fire was upon arrival, Lieutenant Mike Halsey deemed it too "precarious" to attempt rooftop ventilation. "Precarious" most nearly means
    A. Easy
    B. Risky
    C. Difficult
    D. Hot

90. Statistics can sometimes be very misleading. If too much is read into a small portion of the total survey, underlying trends (whatever the study may involve) can become "skewed." "Skewed" most nearly means
    A. Complicated
    B. Simplified
    C. Too focused
    D. Distorted

91. Which of the following statements is true with respect to the drivetrain illustrated below?

A. Gears 2 and 8 will turn clockwise if the first gear turns clockwise.
B. Gears 4 and 5 will turn counterclockwise if gear 2 turns clockwise.
C. Gears 3, 5, and 6 will turn the same direction that gear 2 turns.
D. Gears 3, 5, and 7 will turn counterclockwise if gear 2 turns clockwise.

92. According to the diagram shown, how much pulling effort at point C would be required to lift the weight?

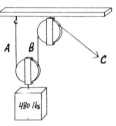

A. 960 pounds   B. 480 pounds

C. 240 pounds   D. 120 pounds

93. The pulley apparatus in question 92 demonstrates what kind of mechanical advantage?
   A. 1:2   B. 2:1   C. 1:1   D. 5:1

94. Which of the following is not considered in the truest form to be an example of a simple machine?
   A. A two-stroke gasoline powered engine
   B. A ramp used to lift heavy objects
   C. A wedge-shaped device used to split wood
   D. A pry bar

95. The Halligan tool illustrated below is another tool available to firefighters to conduct a forcible entry. If the handle portion measures 28 inches in length and the distance from the base of the handle to the tip of the pointed end measures 8 inches in length, how much effort is required at the end of the handle to pry off a door lock shackle that offers 575 pounds of pry resistance?

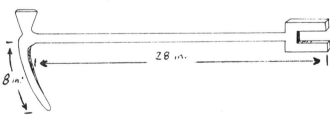

A. 244.2 pounds of force
B. 178.9 pounds of force
C. 164.3 pounds of force
D. 152.7 pounds of force

96. Referring to question 95, what kind of mechanical advantage is gained by using the Halligan tool on the lock described?
   A. 2.35 : 1   B. 3.21 : 1   C. 3.5 : 1   D. 3.76 : 1

COMPLETE FIRE-FIGHTERS EXAM PREPARATION BOOK

97. Three firefighters attempted to lift an aluminum ladder that extended 25 feet in length and weighed 75 pounds. If one of the firefighters served to anchor one end of the ladder to the ground while the other two lifted the other end of the ladder at a point 10 feet from the anchorman, how much resistance would the ladder present to the two doing the actual lifting?
    A. 37.5 pounds of resistance
    B. 75 pounds of resistance
    C. 127.8 pounds of resistance
    D. 93.75 pounds of resistance

98. Which of the statements below is true with respect to the manner in which the three firefighters attempted to lift the ladder as described in the previous question?
    A. Using leverage in this fashion makes it significantly easier for the firefighters to lift the ladder.
    B. The mechanical advantage gained through this kind of leverage is at least 2:1.
    C. Leverage, under the circumstances, is actually working against the two firefighters attempting the lift, rather than serving to their advantage.
    D. The ladder would have been harder to lift at a point farther away from the anchorman.

ANSWER QUESTIONS 99 AND 100 ON THE BASIS OF THE DIAGRAM BELOW

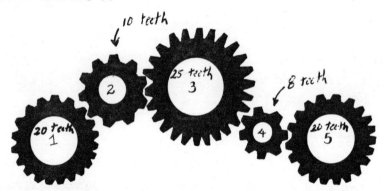

99. If gear 1 was turning at a speed of 400 rpm, how fast would gear 2 turn?
    A. The same speed
    B. Twice as fast
    C. Three times as fast
    D. 20% faster

100. Which of the statements below is true?
    A. Gears 1, 3, and 5 will turn the same direction and gears 1 and 5 will turn at equivalent speeds.
    B. Gears 2 and 4 will turn in the same direction at the same speed.
    C. Gears 1, 2, and 3 will turn clockwise if gears 3 and 4 turn counterclockwise.
    D. Gear 3 will turn the fastest of those shown in the drivetrain.

PRACTICE EXAM 2

ANSWER SHEET TO EXAM 2

1. Ⓐ Ⓑ Ⓒ Ⓓ
2. Ⓐ Ⓑ Ⓒ Ⓓ
3. Ⓐ Ⓑ Ⓒ Ⓓ
4. Ⓐ Ⓑ Ⓒ Ⓓ
5. Ⓐ Ⓑ Ⓒ Ⓓ
6. Ⓐ Ⓑ Ⓒ Ⓓ
7. Ⓐ Ⓑ Ⓒ Ⓓ
8. Ⓐ Ⓑ Ⓒ Ⓓ
9. Ⓐ Ⓑ Ⓒ Ⓓ
10. Ⓐ Ⓑ Ⓒ Ⓓ
11. Ⓐ Ⓑ Ⓒ Ⓓ
12. Ⓐ Ⓑ Ⓒ Ⓓ
13. Ⓐ Ⓑ Ⓒ Ⓓ
14. Ⓐ Ⓑ Ⓒ Ⓓ
15. Ⓐ Ⓑ Ⓒ Ⓓ
16. Ⓐ Ⓑ Ⓒ Ⓓ
17. Ⓐ Ⓑ Ⓒ Ⓓ
18. Ⓐ Ⓑ Ⓒ Ⓓ
19. Ⓐ Ⓑ Ⓒ Ⓓ
20. Ⓐ Ⓑ Ⓒ Ⓓ
21. Ⓐ Ⓑ Ⓒ Ⓓ
22. Ⓐ Ⓑ Ⓒ Ⓓ
23. Ⓐ Ⓑ Ⓒ Ⓓ
24. Ⓐ Ⓑ Ⓒ Ⓓ
25. Ⓐ Ⓑ Ⓒ Ⓓ
26. Ⓐ Ⓑ Ⓒ Ⓓ
27. Ⓐ Ⓑ Ⓒ Ⓓ
28. Ⓐ Ⓑ Ⓒ Ⓓ
29. Ⓐ Ⓑ Ⓒ Ⓓ
30. Ⓐ Ⓑ Ⓒ Ⓓ
31. Ⓐ Ⓑ Ⓒ Ⓓ
32. Ⓐ Ⓑ Ⓒ Ⓓ
33. Ⓐ Ⓑ Ⓒ Ⓓ
34. Ⓐ Ⓑ Ⓒ Ⓓ
35. Ⓐ Ⓑ Ⓒ Ⓓ
36. Ⓐ Ⓑ Ⓒ Ⓓ
37. Ⓐ Ⓑ Ⓒ Ⓓ
38. Ⓐ Ⓑ Ⓒ Ⓓ
39. Ⓐ Ⓑ Ⓒ Ⓓ
40. Ⓐ Ⓑ Ⓒ Ⓓ
41. Ⓐ Ⓑ Ⓒ Ⓓ
42. Ⓐ Ⓑ Ⓒ Ⓓ
43. Ⓐ Ⓑ Ⓒ Ⓓ
44. Ⓐ Ⓑ Ⓒ Ⓓ
45. Ⓐ Ⓑ Ⓒ Ⓓ
46. Ⓐ Ⓑ Ⓒ Ⓓ
47. Ⓐ Ⓑ Ⓒ Ⓓ
48. Ⓐ Ⓑ Ⓒ Ⓓ
49. Ⓐ Ⓑ Ⓒ Ⓓ
50. Ⓐ Ⓑ Ⓒ Ⓓ
51. Ⓐ Ⓑ Ⓒ Ⓓ
52. Ⓐ Ⓑ Ⓒ Ⓓ
53. Ⓐ Ⓑ Ⓒ Ⓓ
54. Ⓐ Ⓑ Ⓒ Ⓓ
55. Ⓐ Ⓑ Ⓒ Ⓓ
56. Ⓐ Ⓑ Ⓒ Ⓓ
57. Ⓐ Ⓑ Ⓒ Ⓓ
58. Ⓐ Ⓑ Ⓒ Ⓓ
59. Ⓐ Ⓑ Ⓒ Ⓓ
60. Ⓐ Ⓑ Ⓒ Ⓓ
61. Ⓐ Ⓑ Ⓒ Ⓓ
62. Ⓐ Ⓑ Ⓒ Ⓓ
63. Ⓐ Ⓑ Ⓒ Ⓓ
64. Ⓐ Ⓑ Ⓒ Ⓓ
65. Ⓐ Ⓑ Ⓒ Ⓓ
66. Ⓐ Ⓑ Ⓒ Ⓓ
67. Ⓐ Ⓑ Ⓒ Ⓓ
68. Ⓐ Ⓑ Ⓒ Ⓓ
69. Ⓐ Ⓑ Ⓒ Ⓓ
70. Ⓐ Ⓑ Ⓒ Ⓓ
71. Ⓐ Ⓑ Ⓒ Ⓓ
72. Ⓐ Ⓑ Ⓒ Ⓓ
73. Ⓐ Ⓑ Ⓒ Ⓓ
74. Ⓐ Ⓑ Ⓒ Ⓓ
75. Ⓐ Ⓑ Ⓒ Ⓓ
76. Ⓐ Ⓑ Ⓒ Ⓓ
77. Ⓐ Ⓑ Ⓒ Ⓓ
78. Ⓐ Ⓑ Ⓒ Ⓓ
79. Ⓐ Ⓑ Ⓒ Ⓓ
80. Ⓐ Ⓑ Ⓒ Ⓓ
81. Ⓐ Ⓑ Ⓒ Ⓓ
82. Ⓐ Ⓑ Ⓒ Ⓓ
83. Ⓐ Ⓑ Ⓒ Ⓓ
84. Ⓐ Ⓑ Ⓒ Ⓓ
85. Ⓐ Ⓑ Ⓒ Ⓓ
86. Ⓐ Ⓑ Ⓒ Ⓓ
87. Ⓐ Ⓑ Ⓒ Ⓓ
88. Ⓐ Ⓑ Ⓒ Ⓓ
89. Ⓐ Ⓑ Ⓒ Ⓓ
90. Ⓐ Ⓑ Ⓒ Ⓓ
91. Ⓐ Ⓑ Ⓒ Ⓓ
92. Ⓐ Ⓑ Ⓒ Ⓓ
93. Ⓐ Ⓑ Ⓒ Ⓓ
94. Ⓐ Ⓑ Ⓒ Ⓓ
95. Ⓐ Ⓑ Ⓒ Ⓓ
96. Ⓐ Ⓑ Ⓒ Ⓓ
97. Ⓐ Ⓑ Ⓒ Ⓓ
98. Ⓐ Ⓑ Ⓒ Ⓓ
99. Ⓐ Ⓑ Ⓒ Ⓓ
100. Ⓐ Ⓑ Ⓒ Ⓓ

ANSWERS TO PRACTICE EXAM 2

Refer back to the sketch for any clarification on questions 1–15.

1. *B*	6. *A*	11. *A*
2. *D*	7. *D*	12. *D*
3. *A*	8. *B*	13. *B*
4. *C*	9. *B*	14. *C*
5. *C*	10. *C*	15. *A*

16. *D.* Any time the flow of water is suddenly stopped, the resulting directional change of the water creates a force to be reckoned with. This force, when trapped within a confined system, can lead to serious damage to fire pumps, hose lines, and water mains, consequently endangering firefighters. In firefighting terms, this kind of problem is referred to as water hammer. It can be avoided if all valves and nozzles are opened slowly

17. *A.* If a person were to stand on the leeward side of a window as it is being broken, there exists the possibility that small glass fragments can be inadvertently blown toward that person. This can be particularly dangerous if some of the debris gets into eyes. Corneal cuts and abrasions, in addition to being painful, can cause blindness. Choice B is a correct procedure because it avoids the possibility of glass falling directly on hands or arms. Choice C is a correct statement. Broken glass can be extremely sharp and can severely cut hands even if gloves are worn. Choice D is a correct procedure as well. Breaking a window from its top down helps prevent the possibility of glass fragments dislodging above and potentially injuring a firefighter attempting to gain entry.

18. *D.* Choice A may seem the correct thing to do; however, if it is unsafe to come to a complete stop, it is better to drive the short distance necessary to get to a point that is safe. Getting off an expressway onto a good shoulder would be one such example. Failure to exercise this caution could result in a far worse accident involving several vehicles. Choice B is incorrect because firefighters should never take it upon themselves to determine whether an accident needs to be reported. All accidents should be reported regardless of the amount of damage sustained to the vehicle(s) involved. Choice C is incorrect as well. Superiors should be notified immediately that an accident has occurred. This is particularly important because if apparatus is detained due to an accident, a replacement unit will need to be dispatched to attend to the emergency. Any further delay than is necessary can lead to tragic results.

19. *A.* When a victim's head is positioned in this manner, it opens the throat for free air passage. The rest of the options are false.

20. *C.* The fire wall in a automobile serves to impede, but not entirely prevent, the spread of flames from the engine to the passenger compartment. Choice A is wrong because the way it is stated, passenger safety is secondary when considering the engine. Engines can always be replaced, but people can't. Choice B is wrong because a fire wall, by its own design, can improve fire safety for passengers, but it cannot totally guarantee their safety. A gasoline vapor explosion can easily breech a car's firewall. When considering choice D, you should be aware

that gas tanks are normally positioned to the rear of a vehicle or at least behind the passenger compartment. If so, passengers are closer to an engine fire than a gas-tank fire. This becomes a secondary worry subordinate to passenger safety.

21. *B.* Air gets progressively thinner with increased altitude. Since air is essential for combustion to take place, increased altitude can diminish an engine's power output. Both choices A and C are false on their own merit. Choice D is a true statement, but it has little bearing on an engine's power output.

22. *D.* When water courses through pipes or hoses, it must overcome certain resistance of friction. Rough interior hose linings, bent couplings, or bends in the hose are major contributors to friction. The effort to overcome these barriers is recognized as pressure loss. Another thing to keep in mind is that the longer a hose lay is, the more friction there is that needs to be overcome. Since choices C and D have three lengths of 40 feet (i.e., 120 feet total) hose versus two lengths (80 feet total) as shown in choices A and B, both of these (C and D) would represent greater amounts of potential friction and resultant pressure loss. Choice D has more bends in its hoselay than C, so D would have additional friction to account for and further loss of pressure.

23. *B.* On the basis of the information given in the answer to question 22, choice B would produce the smallest amount of friction. There are only two 40-foot sections of hose involved and there are only a couple of bends in the hose lay. Consequently, pressure loss would be minimal when compared to the other choices.

24. *C.* If the victim can assist by holding onto a rope, this would be the safest means of pulling the victim to safety. However, if direct assistance is required a ladder can serve as a bridge to safety. In effect, when weight is placed on a ladder that is laid across ice, it will spread that weight over a wider area. This would not eliminate the possibility of the ice breaking any further, but it does allow for a safer means of directly pulling the victim to safety. Choice A may be true, but that is not the reason it would be considered safe when conducting ice rescue. Choice B can almost guarantee that the rescuer would be in the same situation as the victim. Ice, especially in an area that is already weak, will in all likelihood not support the direct weight of someone standing at the edge. Choice D is false if the ice has any thickness.

25. *A.* 100 square feet or 10' x 10'. The question does not ask how much area in square feet could this tarp cover. Rather, it asks how large of a square area could be covered by the tarp in question. The width of the tarp (10 feet) is the deciding factor since it is the smallest dimension of the salvage cover. Since a square's sides must be equal in length, it must not exceed the width of the salvage tarp. Therefore, a 10' x 10' square area could sufficiently be covered by a 10' x 15' salvage tarp.

26. *B.* Water damage from firefighting can be just as bad as the fire itself. Ineffective and excessive use of a hose stream can significantly encumber salvage operations. The action by the firefighter controlling the nozzle can make a salvage operation manageable for a few, or it can be made extensive requiring assistance of many. The other choices either require the effort of more people or are virtually impractical in most situations.

27. *C.* Hospital and nursing home fires are the worst. The safe evacuation of all residents can be an insurmountable problem considering how little time it takes for a fire to spread. If the occupants in a building cannot escape without assistance, the manpower available to evacuate everyone safely may be far under what is actually required. For this reason, the death toll from these kinds of fires is high in comparison to others.

28. *D.* There are two reasons that air should be bled out of hose lines prior to entering a burning

ANSWERS TO PRACTICE EXAM 2

structure. Choice D explains one reason. A second reason is that air expelled from a hose line can potentially ignite an oxygen-starved fuel or fan the flames of an existing fire. Both of these circumstances should be avoided. The remaining choices A, B, and C are incorrect assumptions.

29. **D.** Choices A, B, and C are advantages gained by using a fog or spray over the application of a solid-stream. However, any time water flow meets with more resistance which is the case with a nozzle emitting only a fine spray, higher discharge pressures are required. This is a distinct disadvantage.

30. **B.** Firefighters usually carry a wide variety of forcible entry tools that can essentially pry a door open. Since the incident involves a suspected gas leak, the last thing you want to do is create a spark which can serve as an ignition source to the gas. This is precisely what Choices A, C, and D would do, thus setting the stage for an explosion.

31. **C.**
$$\frac{40\tfrac{5}{8} \text{ revolutions}}{2\tfrac{1}{2} \text{ minutes}} = \frac{X \text{ revolutions}}{1 \text{ minute}}$$

$$2\tfrac{1}{2} X = 40\text{-}5/8$$

$$X = \frac{40\tfrac{5}{8}}{2\tfrac{1}{2}} \text{ or } X = \frac{325}{8} \times \frac{2}{5} \quad \text{(Reciprocal of) } \tfrac{5}{2}$$

$$X = \frac{650}{40} = 16\tfrac{10}{40} = 16\tfrac{1}{4} \text{ (reduced)}$$

Therefore the shaft turns at $16\tfrac{1}{4}$ rpm.

32. **A.** The LCD for the 3 fractions is in thousandths.

Therefore $1\tfrac{3}{10} = \tfrac{13}{10} = \tfrac{1300}{1000}$
$2\tfrac{12}{100} = \tfrac{212}{100} = \tfrac{2120}{1000}$
$1\tfrac{17}{1000} = \tfrac{1017}{1000}$

$$\frac{1300}{1000} + \frac{2120}{1000} + \frac{1017}{1000} = \frac{4437}{1000} = 4\tfrac{437}{1000} \text{ amps}$$

33. **D.** The perimeter length of a trapezoid is found by adding the length of its four sides.

$2\tfrac{3}{4}$ meter + $\tfrac{2}{3}$ meter + $1\tfrac{7}{8}$ meter + 1 meter

The LCD for the three fractions is 24.
$2\tfrac{3}{4}$ or $\tfrac{11}{4} = \tfrac{66}{24}$
$\tfrac{2}{3} = \tfrac{16}{24}$
$1\tfrac{7}{8}$ or $\tfrac{15}{8} = \tfrac{45}{24}$
$1 = \tfrac{24}{24}$
$\tfrac{66}{24} + \tfrac{16}{24} + \tfrac{45}{24} + \tfrac{24}{24} = \tfrac{151}{24} = 6\tfrac{7}{24}$

Therefore, the perimeter length of the trapezoid shown is $6\tfrac{7}{24}$ meters.

34. C. To figure the booster line's external diameter, the hose wall's thickness must be counted twice.

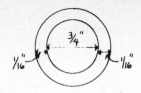

Therefore, ¹⁄₁₆ + ¹⁄₁₆ + ¾ = X diameter
The LCD is 16.
¹⁄₁₆ + ¹⁄₁₆ + ¹²⁄₁₆ = ¹⁴⁄₁₆ or ⅞ inch outside diameter.

35. B. A centimeter, as the name implies, means ¹⁄₁₀₀ of a meter. Another way to look at it is that there are 100 centimeters in a meter. Therefore,

$$2.67 \text{ meters} \times \frac{100 \text{ cm}}{1 \text{ meter}} = 267 \text{ cm}$$

36. D. The thickness of the coupling can be determined by the external diameter minus the internal diameter.

$$2⅞ - 2⅓ = X$$
$$2³⁄₈ - ⅓ = X$$

The LCD = 24. ⅞ = ²¹⁄₂₄ and ⅓ = ⁸⁄₂₄
Therefore, 2²¹⁄₂₄ - 2⁸⁄₂₄ = ¹³⁄₂₄ which is .542 inches. That represents the coupling thickness for both sides. Dividing by 2 will yield .271, or the thickness of one wall.

37. B. If $R = E/I$, we simply multiply both sides of the equation by I to determine E.

$$R \times I = E$$

38. D. If we divide the heat transformed into work by the total amount of heat produced and multiply that quotient by 100, we can determine this particular engine's thermal efficiency.

$$\frac{60,000 \text{ BTU}}{72,500 \text{ BTU}} \times 100 = X\% \text{ thermal efficiency}$$

.8275 x 100 = X%
X = 82.75%

39. B. By multiplying the total BTU produced by the engine to its thermal efficiency, we can determine the number of BTU that are utilized for work.

87,500 BTU x .613 = 53,637.5 BTU

40. C. The circumference of the smaller pulley is insignificant, since we are already given the ratio that represents the pulley dimensions. If we set this up as a direct proportion, it would be as follows:

$$\frac{3}{1} = \frac{X \text{ rpm}}{300 \text{ rpm}} \quad X = 900 \text{ rpm}$$

However, this is an incorrect answer because we know that it requires the smaller pulley to turn three times to turn the larger pulley once. This is the principle of torque which was discussed in the mechanical principles section of this book. If we use an inverse proportion, we can correctly solve this problem.

$$\frac{1}{3} = \frac{X \text{ rpm}}{300 \text{ rpm}} \quad 3X = 300 \text{ rpm}; \ X = 100 \text{ rpm}$$

41. *C.* The circumference of a circle is equal to

   Diameter x π

   Therefore,

   19.25 = Diameter x (3.1416)

   $$\frac{19.25}{3.1416} = X; \ X = 6.127 \text{ inches diameter.}$$

42. *B.* Viewing a ladder against the wall of a building should remind you of a right triangle. The wall of the building is perpendicular to the ground. We can use the Pythagorean theorem to solve this problem. Six feet represents the base of the right triangle and the ladder itself can be considered the hypotenuse. The unknown variable is the other side of the triangle. The theorem states that

   $$A^2 + B^2 = C^2$$

   (Of course, "squaring" a number is multiplying it by itself. $2^2 = 4$.) In this case,

   $$6^2 + X^2 = 25^2, \text{ or}$$

   $$36 + X^2 = 625.$$
   $$X^2 = 589$$
   $$X = 24.27.$$

   So the ladder reaches 24.27 feet.

43. *C.* The above the ground area of a circle is determined by multiplying π by the square of the circle's radius. Therefore,

   π x $(7.35)^2$ = the area of the circle.
   3.1416 x 54.0225 = X
   X = 169.72 square feet.

   Choice A represents the circumference of this circle, not its area.

44. *B.* First, it is necessary to determine the area of the truss (i.e., triangle). Since the area of a triangle = (½) Base x Height, the truss has an area equal to

   (½)(22.5 feet)(12 feet) = 135 square feet.

   The volume of space afforded by a roof of

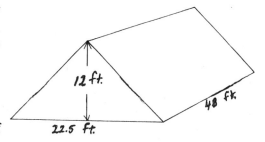

219

these dimensions is equal to the area of the truss (i.e., triangle) x the length of the roof. Therefore, this roof can provide 135 sq. ft x 48 feet or 6,480 cubic feet of space that can be used for storage.

45. *B.* The original 14-liter solution was 35% benzene and 65% water. Therefore, there was 14 x .35 or 4.9 liters of benzene and 14 x .65 or 9.1 liters of water. If we add 4.5 liters of water to this solution, we will then end up with 4.9 liters of benzene in 18.5 liters of solution. To figure the new percentage of benzene present in solution, we divide the number of liters of benzene by the total volume of solution and multiply that quotient by 100.

$$\frac{4.9 \text{ liters of benzene}}{18.5 \text{ liters of solution}} \times 100 = 26.49\%$$

46. *A.* $(X + 3)(X - 8) = X^2 - 8X + 3X - 24 = X^2 - 5X - 24.$

47. *D.* Static electricity and flammable material by themselves pose no problem. However, if the two are combined as stated in choice D, a real problem can result. Choice C can pose that kind of a problem for an individual; however, it was not discussed in the passage.

48. *C.* Anytime an open flame is in close proximity to a flammable substance there exists the possibility of ignition. Air humidification was noted in the reading as being incompatible for some industries but nonetheless is used by some businesses.

49. *A.* Only choice A is true.

50. *B.* Since copper is an excellent conductor and it interconnects all the tanks, we can assume the company involved is utilizing bonding as a means of controlling static electricity.

51. *D.* All the options given in the question are conditions that contribute to a static electric spark.

52. *B.* Ionizers used in various industries siphon off static charge that can accumulate on equipment surfaces. It does not accomplish what is stated in choices A and C. Choice D may seem to be right, but static electricity is not normally thought of as being life threatening by itself. Usually discomfort is experienced by an individual who inadvertently receives a static electrical shock. The real danger that exists for an individual occurs if this spark were to occur in an environment that has flammable vapors present; an explosion would likely result.

53. *C.* Choice C best describes the article as a whole. The remainder of choices do, in fact, refer to portions of the article, but fail to encompass the basic meaning of the passage.

54. *D.* Boiling liquid expanding vapor explosion or, in abbreviated form, a BLEVE. Choice A is a trick option that misspells the true abbreviated term BLEVE. This kind of question is one of the main reasons it is important to read and consider all the choices given. It is very easy to prematurely conclude that the first choice read is correct and ignore the remaining alternatives. Choices B and C are items discussed; however, neither in itself is a technical term.

55. *A.* Vaporization. Condensation is a process exactly the opposite of vaporization. A vapor is reduced or condensed to its original liquid form, which always demonstrates a higher density than the gas. The consequence of this process would be a reduction of internal pressure, not an increase. Pyrolysis is the conversion of a solid material to a vapor state prior to combustion. This process does not apply to liquids.

56. *B.* The vacuum created by a liquid being withdrawn from a closed container will cause it to col-

ANSWERS TO PRACTICE EXAM 2

lapse at a certain point. Choice A could be right if the vacuum created did not exceed the tank's structural limitations. Choice D is a possibility as well, providing the pump's capacity is small and the tank's integrity is maintained. If the vacuum created cannot be overcome by a small pump, it is a distinct possibility the pump will cease to function. Choice C is wrong because it is not an expansion that occurs within a tank when its contents are being withdrawn. Rather, it is a contraction. Choice B, as it is stated, best sums up what may happen without any other contingencies made known.

57. *C.* With the vents directed in a vertical manner as shown in Choice C, the vapor release if ignited would not affect the neighboring storage tank. This kind of impingement is demonstrated by choices A and B. Choice D is not practical because the vent would be obstructed by the liquid itself and serve only as an open stopcock to drain the liquid as pressure is increased by vaporization.

58. *A.* Routine inspection was suggested; however, there was no clarification of the timetable involved in the reading.

59. *C.* This reading was primarily concerned with how to eliminate the potential dangers that dust can pose in a work environment. Choices B and D were mentioned in the article but neither conveyed the general meaning of the article. Choice A is wrong by its own merit.

60. *D.* Choices A through C each contribute to a safer means of operating a ventilation fan in a dusty environment.

61. *C.* The key words in this question are most direct. The other options are, in fact, methods for enhancing safety. However, the most direct means of coping with a dust problem is to contain the dust as best as possible at its point of origin. Regular maintenance and cleanup can then be focused on those confined areas rather than controlling dust at all levels of a given building.

62. *A.* This would be the worst method of cleaning up accumulated sawdust. Since sawdust is a combustible product, it becomes much more so if suspended in air. The finer the sawdust, the longer it will remain suspended in air, thus exacerbating the potential danger. Even though brooming sawdust into piles creates a certain degree of dust suspension, it does not compare to what air blowers render. The other two options mentioned are considered safer ways to handle sawdust cleanup.

63. *D.* This choice best explains why air-suspended combustible particles are more susceptible to flammability. Choice A is a true statement by itself, but it is not the true reason for susceptibility. Choice B and C are blatantly wrong.

64. *C.* The hospital is northwest of the fire department, according to the directional legend provided.

65. *A.* Since the victim collapsed on the northeast side of Ester Fairway, that would place him closer to Bristol and King Street than the Clubhouse. Remember, time is of the essence in any kind of emergency. If rescue workers have to carry most of their equipment across the Fairway just to reach the victim, precious seconds are lost. This is the primary reason choice B is incorrect. Both choices C and D are wrong because both sets of directions run counter to one-way streets. You cannot go northwest on Vincent or southwest on 35th Avenue.

66. *D.* Northeast. If the question had asked what direction the hospital was in relation to the pharmacy, choice C would have been the correct answer.

67. *B.* According to the diagram's legend, blocks displaying a square or rectangular symmetry have

COMPLETE FIRE-FIGHTERS EXAM PREPARATION BOOK

fire hydrants located in the eastern corner. Only choice B describes a fire hydrant location that actually exists.

68. *D.* Choices A, B, and C are true. Choice D is wrong because Cambridge Court is actually south of the Ester Fairway Clubhouse.

69. *B.*

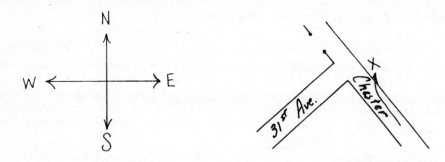

Under the circumstances given, it should be fairly obvious that the commuter was heading in a northwesterly direction.

70. *C.* Choice A is wrong because you will not intersect Chester proceeding southwest on 19th Avenue from the fire station. This direction will take you away from Chester. Choice B is incorrect because it runs counter to a one-way street. Choice D may seem the correct answer, but remember that the accident took place half a block southeast of 31st Avenue. To attempt rescue from this direction would necessitate a time-wasting sharp right turn on Chester from Hazelwood to go that extra half a block. Choice C is the most direct route to take.

71. *D.* Choices A and B are incorrect because, according to the diagram's "facts to consider," rail traffic effectively blocks Chester and Bristol Streets each hour between 6 a.m. and 7 p.m. Choice C is not possible because another fact to consider is that the intersection of Beach Point and 25th Avenue is closed for construction purposes between 10 a.m. and 5:30 p.m. This effectively closes the Beach Point route to the hospital. Choice D may seem to be a roundabout way of getting to the hospital; however, it is the only alternative left, next to waiting for the train to pass. Considering the fact that the victim's breathing is labored, it is best to proceed to the hospital following Choice D instead of waiting at a railroad crossing.

72. *C.* Choice A would have been correct had it said the second house on the right. From the direction this person is coming from, the southwest side of Bristol would be to the right, as described in choice C. Choice B would have been the correct choice had this individual elected to drive instead of walking.

73. *C.* One of the facts to consider is that the Ester River flows in a southeasterly direction. Even though the rescue effort was expedient thanks to an alert citizen, the teenagers would probably have been swept past the 31st Avenue bridge by the time the authorities had arrived. Therefore, considering the river's current, it would probably be best to concentrate rescue efforts below the railroad trestle and work their way up the river from there. If more than a couple of minutes had elapsed from the time the teenagers were spotted to the time that rescuers responded, then the 16th Avenue bridge probably would have been a good point to start a rescue operation.

74. *A.* Northwest. Choice C would have been correct if the question had been worded to ask where the ECR Inc. building is in relation to the post office.

75. D. 2627 Chester.

This illustration shows fire hydrant location on triangular blocks, according to the "facts to consider" in the diagram.

76. B. Choice A would have been correct if the question had specified that entrance as the destination. However, it is the northeast entrance to the post office that is of concern in this question. Choice C is incorrect because you cannot proceed in a southeasterly direction from Cambridge Court and expect to intersect 31st Avenue. Choice D is incorrect because you cannot reach 28th Avenue from Piedmont directly. Choice B is the most direct route offered.

77. C. Inside calipers would be the most appropriate tool to use. Choice A is an illustration of outside calipers that can be used to compare the outside diameter of two different pipes. Choice B is a protractor, which is used more appropriately to draw or plot angles on a given surface, not to measure various diameters. Choice D can roughly measure internal and external diameters, but it cannot offer the same degree of accuracy that is afforded by a calipers.

78. A. The instrument shown is a depth gauge. Its principle use is to determine the depth of holes, mortices or grooves.

79. D. A try square and carpenter's framing square are both designed only to determine the true squaring of a given angle (i.e., 90°). A combination square, however, has the versatile feature of determining a miter square.

80. B. A Torx screwdriver is needed for this screw.

81. D. The socket wrench has the added feature of a ratchet handle which allows for turning bolts in tight areas. Choices A and B would not be practical under the circumstances given. Choice C is a possibility, but since most lack a ratchet mechanism, it requires more time and effort to loosen the bolt. It is not nearly as convenient as a socket wrench.

82. A. Only Allen wrenches have a design of this nature. They are used to tighten or loosen machine setscrews (see example).

83. C. A triangular file will serve the purposes described in the question. Choice D is incorrect because the question referred to files only, not electrical tools.

NOTE: Even though vocabulary tests are rarely seen on firefighter exams, that does not preclude the possibility of encountering them. A few questions have been provided to test your skill in this area. Remember, how a word is used in context or its derivation can help you to discern the meaning of an unknown term.

84. B. A conflagration is essentially a fire that gets out of control and consequently inflicts terrific amounts of damage. The factors given in the question that are attributed to the development of a conflagration make it fairly obvious that conditions are ripe for an out-of-control fire.

Lateral and vertical fire extension through radiant and convective heat transmission would make short order of such situations.

85. *B.* Acrophobia means a fear of heights. If how the word was used in the context of the question did not offer enough of a clue as to its meaning, the etymology table provided in this book would have helped. "Acro-" means furthest or highest point and "phobia" means "fear of."

86. *D.* Ineptitude means lack of aptitude or awkwardness.

87. *B.* Dilapidated most nearly means deteriorated.

88. *C.* If a metal workpiece is said to be malleable, that means it has the capability of being shaped either by a hammer or under some other form of applied pressure. An excellent example of something that is malleable is a rivet whose head is rounded by a ball peen hammer.

89. *B.* Precarious most nearly means risky.

90. *D.* Skewed most nearly means distorted. If too much emphasis is placed on only a portion of a survey, the conclusions that may be drawn from the study may distort otherwise obvious trends or directions.

91. *D.* Gears 3, 5, and 7 will turn counterclockwise if gear 2 turns clockwise. As a rule of thumb, if gears are directly aligned end to end as shown in the illustration, the odd-numbered gears in the sequence will always turn in the same direction while the even-numbered gears will always turn the opposite direction.

92. *C.* Since cables A and B both act to support equally the weight being lifted, the load factor is evenly distributed between the two. In other words, cable A supports 240 pounds and cable B supports the other 240 pounds. Therein lies the mechanical advantage gained by using this pulley system.

93. *B.* 2:1, because it requires only 240 pounds of effort to lift the weight depicted versus 480 pounds if it were lifted directly. The lifting advantage gained by using the pulley is doubled to 2:1. Choices A, C, and D are not reflective of any mechanical advantage.

94. *A.* Choice B is an example of an inclined plane. Choice C is a wedge and Choice D is an example of a lever. All of which are classified as simple machines.

95. *C.* To calculate the effort required,

$$\text{Effort} = \frac{\text{Resistance} \times \text{Resistance Distance}}{\text{Effort Distance}}$$

Therefore,

$$E \quad \frac{575 \text{ lbs} \times 8''}{28''} = \frac{4600}{28} = 164.3 \text{ pounds of force}$$

96. *C.* Since it requires only 164.3 pounds of force to overcome 575 pounds of pry resistance, the Halligan tool offers a 3.5:1 mechanical advantage of prying the lock off the door versus pulling on it directly (i.e., 575 lbs / 164.3 lbs).

97. **D.** The anchorman described in the question essentially serves as the fulcrum. Half the length of the ladder is considered the resistance distance and 10 feet represents the effort distance. Therefore,

$$\text{Effort} = \frac{75 \text{ pounds} \times 12.5 \text{ feet}}{10 \text{ feet}} = \frac{93.50}{10} \text{ or } 93.75 \text{ pounds}$$

of resistance would be presented to the two firefighters attempting lift under the circumstances described.

98. **C.** This is the only true statement given.

99. **B.** For every one revolution made by gear 1, gear 2 turns twice. Therefore, if gear 1 is turning at 400 rpm, gear 2 is turning at 800 rpm.

100. **A.** Every other gear in a drivetrain of this design will turn in the same directions. Since gears 1 and 5 have the same number of teeth, both will turn at equivalent speeds. Choice B is false only because gear 4 will turn faster than gear 2 because of its fewer teeth. Choice C is incorrect because gears 1 and 2 turn opposite to one another as do gears 3 and 4. Choice D is wrong because the larger the gear (i.e., the more teeth present), the slower it will turn, providing the driver gear size remains constant. This is an example where speed is decreased there is a corresponding or proportional increase in torque.

TEST RATINGS ARE AS FOLLOWS:

95–100 correct	EXCELLENT
87–94 correct	VERY GOOD
81–86 correct	GOOD
75–80 correct	FAIR
74 or below correct	UNSATISFACTORY

Go back to the questions you missed and determine if the question was just misinterpreted for one reason or another, or if it reflects a particular weakness in subject matter. If it is a matter of misinterpretation, try reading the question more slowly while paying close attention to key words such as *not, least, except, without,* etc. If, on the other hand, you determine a weakness in a particular area, don't despair. That is what this book is for; to identify any area of weakness before you take the actual exam. Reread the material on the area of concern in this book. If you still feel a need for supplemental material, your local library is an excellent source.

PRACTICE EXAM 3

Time allotted: 2 hours

DIRECTIONS: Each question has four choices for answers, lettered A, B, C, and D. Chose the best answer and then, on the answer sheet provided, find the corresponding question number and darken, with a soft pencil, the circle corresponding to the answer you have selected.

INSTRUCTIONS FOR QUESTIONS 1–17

Study the following diagram for only 5 minutes. When your time is over, turn to the next page to answer questions 1 through 17 relating to the diagram you have studied. YOU MAY NOT LOOK BACK AT THE DIAGRAM. NOTES ARE PROHIBITED. REMEMBER: This is a memory exercise, so any review of this diagram is prohibited on the actual examination. Thus, any review during the practice exam will skew your test results.

DIAGRAM
The diagram depicts one kind of mechanical format for a pump control panel.

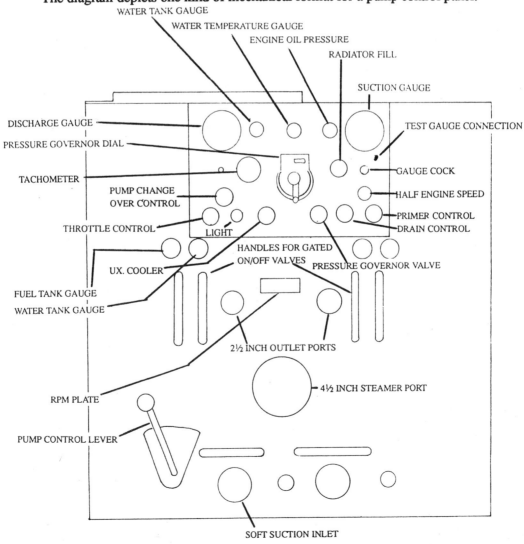

PRACTICE EXAM 3

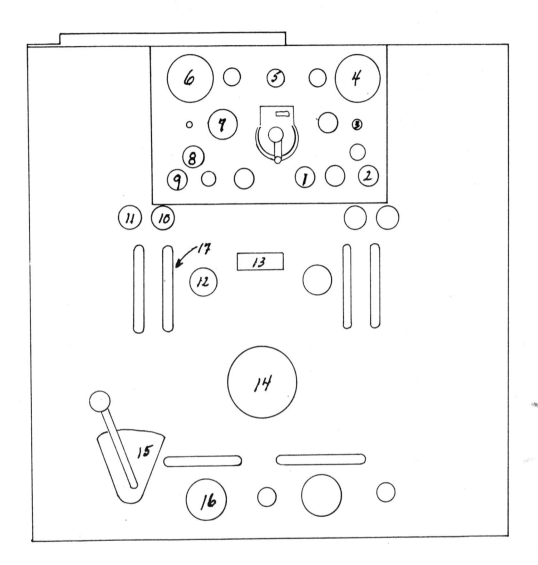

ANSWER QUESTIONS 1 THROUGH 17 ON THE BASIS OF THE DIAGRAM ON PAGE 229 (which you may consult)

1. What aspect of the pump panel is indicated by number 1?
   A. Drain control
   B. Primer control
   C. Throttle control
   D. Pressure governor valve

2. What aspect of the pump panel is indicated by number 2?
   A. Suction gauge
   B. Primer control
   C. Drain control
   D. Pressure governor valve

3. What aspect of the pump panel is indicated by number 3?
   A. Gauge cock
   B. Half engine speed
   C. Test gauge connection
   D. Drain control

4. What aspect of the pump panel is indicated by number 4?
   A. Discharge gauge
   B. Tachometer
   C. Suction gauge
   D. Handles for gated valves

5. What aspect of the pump panel is indicated by number 13?
   A. 2½-inch outlet ports
   B. Handle for gated valves
   C. Rpm plate
   D. Drain control

6. What aspect of the pump panel is indicated by number 7?
   A. Pressure governor dial
   B. Test gauge connection
   C. Water-tank gauge
   D. Tachometer

7. What aspect of the pump panel is indicated by number 5?
   A. Engine oil pressure
   B. Water-tank gauge
   C. Tachometer
   D. Water-temperature gauge

8. What aspect of the pump panel is indicated by number 6?
    A. Suction gauge
    B. Discharge gauge
    C. Engine oil pressure
    D. Pump changeover control

9. What aspect of the pump panel is indicated by number 8?
    A. Pump changeover control
    B. Throttle control
    C. Auxiliary cooler
    D. Half engine speed

10. What aspect of the pump panel is indicated by number 12?
    A. rpm plate
    B. Soft suction inlets
    C. 2½-inch outlet port
    D. Primer control

11. What aspect of the pump panel is indicated by number 9?
    A. Throttle control
    B. Light
    C. Auxiliary cooler
    D. Pump changeover control

12. What aspect of the pump panel is indicated by number 11?
    A. Water tank gauge
    B. Drain control
    C. Fuel tank gauge
    D. Engine oil pressure

13. What aspect of the pump panel is indicated by number 10?
    A. Fuel tank gauge
    B. Water tank gauge
    C. Suction gauge
    D. Rpm plate

14. What aspect of the pump panel is indicated by number 14?
    A. 4-inch steamer port
    B. Soft suction inlet
    C. 4½-inch outlet port
    D. 4½-inch steamer port

15. What aspect of the pump panel is indicated by number 16?
    A. 2½-inch steamer port
    B. Suction gauge
    C. Discharge gauge
    D. Soft suction inlet

16. What aspect of the pump panel is indicated by number 15?
    A. Handle for a gated valve
    B. Pump control lever
    C. Throttle control
    D. Primer control

17. What aspect of the pump panel is indicated by number 17?
    A. Pump control lever
    B. Pressure governor dial
    C. Handle for a gated valve
    D. rpm plate

18. This diagram shows a side perspective of a two-bedroom mobile home with a flat roof. The fire is in the room to the left, a bedroom; the room to the right is the living room. According to the illustration, points A, B, C, and D represent possible wall or ceiling breaches for ventilation. Which of these points would be the correct place to create an opening to allow trapped heat and smoke to escape?

    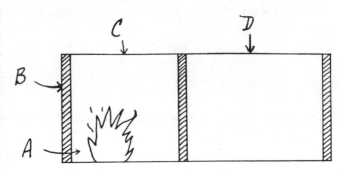

    A. Point A   B. Point B   C. Point C   D. Point D

19. Several hundred yards upwind from a tank-train derailment, a fire lieutenant examines the accident through binoculars. What is immediately apparent is a white placard that highlights a black skull and crossbones. Without knowing precisely what kind of chemical is in the overturned tank, what can be assumed about the contents?
    A. It is an inert substance that does not warrant concern.
    B. It is some form of poison dangerous to people.
    C. It is dangerous only with respect to what comes into direct contact with it.
    D. It is going to involve only routine cleanup of the accident site.

20. A firefighter on the roof of a three-story structure tosses one end of a rope to co-workers waiting below. It is understood that a ventilation fan is needed on top, so the rope's end is tied around the top of the unit. Before the fan is hoisted, a second line is tied to the underside of the unit. As the fan is lifted, the underline is kept semitaut at a position 20 feet from the base of the building. The reason for this practice is:
    A. To assist in lifting the ventilation fan to the top floor.
    B. To prevent the unit from falling.
    C. To keep the fan from twisting.
    D. To keep the equipment from striking the side of the building.

21.

Ordinary Combustibles   Flammable liquids   Electrical Equipment

The diagrams above represent the three different classes of fire. If these three symbols were on a fire extinguisher, which of the following statements would be true?
   A. This fire extinguisher is very versatile with respect to what it can extinguish.
   B. This fire extinguisher can put out only flammable liquid and electrical fires.
   C. This fire extinguisher is capable of extinguishing only ordinary combustible fires.
   D. This fire extinguisher could be used occasionally on electrical fires if caution is exercised.

22. By looking at the same diagram as in question 21, what one factor should be evident about the contents of this extinguisher?
   A. It can be sprayed long distances.
   B. It is hazardous to your health.
   C. It must have an alcohol base.
   D. It must have conductive properties.

23. When hose lines are laid by an apparatus (i.e., a pumper or tanker truck), what is the correct means for a firefighter to anchor the hose before the apparatus moves forward to pay out its load?
   A. Stand atop the hose or coupling.
   B. Wrap the hose twice around the waist.
   C. Face the opposite direction the apparatus is moving and lean forward while straddling the hose.
   D. Grasp the hose firmly and lean slightly backward.

24. When the floor supports for any occupancy weaken because the load factor exceeds capacity, one of three different kinds of collapse can occur. If a floor collapses toward the middle, the event is referred to as a V-type collapse. If one side of a floor support gives way while the other side basically remains intact, that situation is referred to as a lean-to collapse. If both sides of a floor collapse simultaneously, the event is referred to as a floor-to-floor or pancake collapse. Civil engineers consider which of these scenarios the worst in terms of victim survival?
   A. Pancake collapse
   B. V-type collapse
   C. Lean-to collapse
   D. Cannot be determined by the information given.

25. In some cases, wood frame wall interiors serve to vertically extend a fire and make it more difficult to locate and extinguish. Walls are not indiscriminately breached by firefighters to learn of a fire's location. Rather, there are some physical clues that can identify hot spots. Of the following options given, which one is the least likely to provide a hint as to the location of a fire within a partition?
   A. Distinct discoloration of wallpaper
   B. The wall becomes concave in appearance
   C. Paint blistering
   D. Portions of the wall feel warm to the touch

COMPLETE FIRE-FIGHTERS EXAM PREPARATION BOOK

ANSWER QUESTIONS 26 & 27 ON THE BASIS OF THE CHART BELOW

Back injuries	432 days
Eye injuries	26 days
Smoke inhalation	37 days
Burns	12 days
Leg and wrist sprains	97 days
Lacerations	30 days

The chart quantifies time loss accidents over a ten-year period for Ladder Company 1281. It shows the nature of injuries involved and the cumulative work time lost as a consequence of each type.

26. According to the information in the chart, how many work days were lost per year due to employees not wearing approved goggles or face shields in the line of duty?
    A. 2.6 days   B. 37 days   C. 3.7 days   D. 26 days

27. Referring to the same chart, how many work days are lost annually due to firefighters not wearing self-contained breathing apparatus when necessary?
    A. 2.6 days   B. 3.0 days   C. 3.7 days   D. 9.7 days

28. Fire hydrants can be either wet-barrel or dry-barrel depending on their valve design. A wet-barrel hydrant possesses a valve or valves (depending on how many outlets are present) on the top side to control the flow of water. Dry barrels, as the name implies, are dry because the valve that controls the flow of water is at the base of the hydrant. What is the seeming benefit of the dry barrel over the wet barrel?
    A. It has less chance of leaking.
    B. They are cheaper to construct.
    C. They can be utilized in freezing climates.
    D. They are safer to operate.

29. Based on the duct work configuration above, if water under pressure enters inlet port C, and manual valves D, F, and J are open, while H is closed, where would the water flow?
    A. Through outlet port G
    B. Through outlet port I
    C. Through outlet port B
    D. It would be equally split between outlet ports K and L.

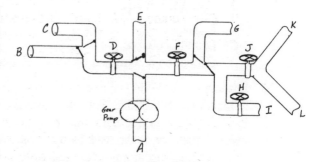

234

30. Referring to the illustration in question 29, if water under pressure enters inlet port B, and manual valves D and H are open, while valves F and J are closed, where would the water flow be directed?
    A. Through outlet port E
    B. Through outlet port G
    C. Through outlet ports K and L
    D. Since there is no place for the water to flow, a static pressure would develop.

31. Referring to the illustration in question 29, if water is suctioned from a tank through inlet port A and only manual valves F and H are open, while D and J are closed, where would the water flow be directed?
    A. Through outlet port I
    B. Through outlet ports K and L
    C. Through outlet port L
    D. Through inlet port E

32. If a firefighter noticed that one of his co-workers was regularly misusing some of the equipment in the line of duty, what would be the best way to handle this problem?
    A. Point out the ineptitude to other co-workers.
    B. Report the infraction to an immediate on-duty supervisor.
    C. Tell the co-worker to straighten up his or her act or get another job.
    D. Advise the co-worker of the proper use of the equipment. Only if the problems persists should a superior become involved in terms of discipline.

33. The progress of a fire is governed by several major factors. These factors are listed below along with other information that has little bearing on fire development.
    1. The nature of the combustible surfaces involved.
    2. The extent of ventilation.
    3. The kind of ignition source involved.
    4. The arrangement of the combustible surfaces involved.
    5. The color of the fuel involved.
    6. The amount of fuel (i.e., exposed combustible surface).
    7. The amount of firefighting personnel on hand.
    8. The amount of light present.

    Choose the one choice that best represents all information within this list that is pertinent to fire progression.
    A. Only numbers 1, 2, 3, and 6 are directly relevant to fire progression.
    B. Only numbers 1, 2, 4, and 6 are directly relevant to fire progression.
    C. Only numbers 2, 3, 5, 7, and 8 are directly relevant to fire progression.
    D. Only numbers 3, 5, 6, and 7 are directly relevant to fire progression.

34. Which of the buildings described below offers the least resistance to fire progression?
    A. A four-story townhouse with an elevator and enclosed stairwells.
    B. A four-story condominium with a centrally located enclosed stairwell.
    C. A four-story hotel with an elevator and centrally located open staircase.
    D. A four-story apartment house with two elevators and enclosed stairwells.

35. One danger that confronts firefighters is the lack of oxygen. It has been demonstrated by various studies that people require a level of about 16% oxygen in the atmosphere to survive. There are a few cases when some people can get by with less, but only for a short time. An open flame will not continue to exist without 16% oxygen either. What kind of correlation can be drawn from this?
    A. The presence of an open flame indicates that there is insufficient air for a firefighter to breath.
    B. Any time an open flame is observed, it is necessary to use a self-contained breathing apparatus.
    C. If a firefighter observes a flame being extinguished for no apparent reason, the breathing apparatus can be taken off.
    D. The presence of an open flame indicates that there is enough oxygen present to sustain life.

36. When uncharged fire hose is being stretched within a building, it is important that it not be inadvertently laid beneath a door or in such a manner that a door may close on the line. What is the main reason for this precaution?
    A. It prevents excessive wear on the hose.
    B. The door may act as an effective clamp on the hose line.
    C. Doors can be damaged by charged lines.
    D. To avoid potential hose rupture.

37. If it is necessary to tap a hole $4/5$ the diameter of the bolt intended to be threaded into it, what would be the recommended tap for a bolt with a diameter of $5/8$ inch?
    A. $3/8$ inch   B. $1/2$ inch   C. $2/3$ inch   D. $3/4$ inch

38. If fresh water has a density of $62 2/5$ pounds per cubic foot and it is known that $2402 1/2$ pounds of water are present in a cylindrical water tank, how many cubic feet of water would that constitute?
    A. 27.1 cubic feet   B. 34.75 cubic feet   C. 38.5 cubic feet   D. 39 cubic feet

39. If one gallon of water weighs $8^{35}/_{100}$ pounds, how many gallons of water would weigh 16,000 pounds?
    A. 1916.17 gallons
    B. 2435.07 gallons
    C. 2,789 gallons
    D. 3,916.07 gallons

40. If one side of a hexagonal nut measured $2/3$ centimeter in length, what would be the length of all sides added together, or in other words, its perimeter length?
    A. 4 cm   B. 3.75 cm   C. 3.12 cm   D. 2.91 cm

41. If the factory specification of a shaft's diameter is 1.75 inches, but was actually determined to be 1.732 inches with a micrometer calipers, what is the difference involved?
    A. .018 inches   B. .0018 inches   C. .18 inches   D. 1.8 inches

42. A fire started by an electrical short was quickly put out with a dry chemical extinguisher. If the extinguisher has a capacity of $8 1/2$ pounds, and $3 1/4$ pounds were left after the fire, what percentage of extinguishing agent was actually used?
    A. 45.92%   B. 49.61%   C. 57.52%   D. 61.76%

ANSWER QUESTIONS 43 THROUGH 45 ON THE BASIS OF THE SURVEY BELOW

	# of fires reported	# of fatalities from fires	# of fires caused by arson
1988			
Fire District 12	137	14	14
Fire District 18	252	25	32
Fire District 22	16	1	1
1989			
Fire District 12	149	7	18
Fire District 18	213	12	47
Fire District 22	27	2	3

43. Fire District 18 is shown to have had fewer fires reported in 1989 than in 1988. This reflects what degree of decrease in terms of percentage?
    A. 18.3%   B. 16.7%   C. 15.5%   D. 13.2%

44. Which fire district experienced the largest percentage increase from 1988 to 1989 in fires attributed to arson?
    A. Fire District 22
    B. Fire District 18
    C. Fire District 12
    D. Cannot be determined from the information given.

45. If Fire District 12 experienced a 450% increase of fatalities related to fires when comparing figures from 1992 to 1988, how many fatalities were there in 1992?
    A. 57   B. 58   C. 63   D. 77

46. Assuming that a hose nozzle weighs 4.75 pounds and is an alloy comprised of 67% copper, how many ounces of copper are present in this hose nozzle?
    A. 3.1825 ounces   B. 60.79 ounces   C. 54.93 ounces   D. 50.92 ounces

47. Compression ratios reflect the changing volume of space between the piston head and the top of the cylinder between when the piston is at bottom dead center and when it is at top dead center. Basically, the higher the compression ratio, the more usable power can be extracted from the fuel involved. If there are 8 cubic inches of space left in a cylinder when the piston is at its top center, and there are 58 cubic inches of space when the piston is at bottom center, how would this best be expressed as a compression ratio?
    A. 8 : 58   B. 58 : 8   C. 7.25 : 1   D. 1 : 6.25

48. If an air sample from an older insulated warehouse was shown to have 675 micrograms of asbestos per liter, how many micrograms of this particulate matter would be in a cubic meter? (Hint: A liter is considered to be a cubic decimeter.)
    A. 0.675 micrograms
    B. 6.75 micrograms
    C. 67.5 micrograms
    D. 675,000 micrograms

49. How much guy wire would be required to support a ham radio antenna if the collar to which the guywire is connected is 23 feet off the ground and the anchor turnbuckles are evenly spaced 15 feet away from its base? (Allow an extra 8 inches for fastening the ends of each of the four wires.)
    A. 112.52 feet   B. 107.5 feet   C. 101.92 feet   D. 95.78 feet

50. An insurance adjuster figured that one room of a home suffered enough smoke and water damage from the fire to warrant complete replacement of all the drywalling. If this room had the dimensions of 25 feet x 27 feet x 9.6 feet, how many square feet of drywalling would be required for the job? (Assume the door and windows account for 48.5 square feet.)
    A. 1673.4 square feet
    B. 1251.4 square feet
    C. 949.9 square feet
    D. 1624.9 square feet

51. If a spherical liquified-Halon storage container measured 72.75 centimeters in diameter, how many cubic centimeters of Halon could this tank maximally accommodate?
    A. 5,542.37 cubic centimeters
    B. 107,541.2 cubic centimeters
    C. 201,603.76 cubic centimeters
    D. 500,944.3 cubic centimeters

ANSWER QUESTIONS 52 THROUGH 54 ON THE BASIS OF THE PASSAGE

Steel-bottle tanks are frequently used to store relatively small amounts of liquified and compressed gases. To release these gases in a controlled manner some form of valve is provided. When pressurized gases are released in an uncontrolled fashion, a violent explosion may result, particularly if the vapor cloud that is formed ignites. There are several causes for tank failure. One possibility involves inadvertent damage to the tank itself such as a puncture or the valve mechanism being broken off. Another possibility, although very remote, may involve combustion of the material within the tank. In this case, an extreme explosion would result. Static electricity may seem the likely suspect; however it normally is an insignificant factor in a closed system such as a pressurized tank. Nevertheless, if a spark or some ignition source is provided within a tank filled with compressed flammable gases, it is inevitable that combustion will occur. Another situation that can lead to tank failure involves heat-inspired gas expansion. If a tank is filled to near its capacity and sufficient heat is applied, the internal pressure can build to a point that exceeds the tank's limitations. Tank rupture is the likely consequence unless a safety valve allows excess pressure to be bled off. The last possibility worth mentioning is the subjection of a tank to an open flame. If the tank's wall absorbs enough heat, its structural integrity may be compromised or, in other words, it becomes weakened. At some point, the tank may no longer contain the pressurized contents, and again an explosion may result.

Some obvious precautions should be exercised any time compressed gas is handled. Tanks should be secured so as to minimize the possibility of a tank leaning over or accidentally rolling. This is one way in which valves can be inadvertently broken off. Typically, tanks have protective caps that should be threaded over the valve. These should always be left on, especially during transport, and not removed until the tank is actually used. Tanks should always be kept in a cool area free from any open flame or potential heat source. These few simple guidelines are important to follow if compressed or liquified gases are to be handled in the safest manner possible.

52. Which of the following choices would be a suitable title for the article just read?
    A. What Not to Do While Handling Bottle Tanks.
    B. The Potential Hazards of Stored Pressurized Gases and How Best to Handle Them Safely.
    C. The Likely Prospect of What Will Happen if Gas Cylinders Are Mishandled.
    D. How Employees Can Best Protect Themselves against the Hazards of Handling Liquified and Compressed Gases.

53. How may reasons for tank failure were discussed in the passage?
    A. 2   B. 3   C. 4   D. 5

54. What one attribute mentioned in the reading renders a tank potentially explosive when it is exposed to moderate amounts of heat?
    A. Thin cylinder walls
    B. Faulty valve
    C. Reactivity of the tank's contents to heat
    D. A nearly full tank

55. Which of the following statements is false according to what was read?
    A. Static electricity is a significant factor to contend with when considering stored gasses (i.e., closed systems).
    B. A compressed gas cylinder subjected to an open flame is likely to fail.
    C. Internal combustion of a tank's contents is an unlikely event.
    D. If enough heat is absorbed by a cylinder's wall, its strength may be sacrificed.

56. On some compressed gas cylinders a safety relief valve is present for what purpose?
    A. Bleed off excess pressure.
    B. To determine how much pressure is present within a tank.
    C. It circumvents any possibility of tank rupture.
    D. It serves as a thermostat.

57. Which of the following statements is true according to safety guidelines mentioned in the article?
    A. Protective caps should be threaded on at all times until the tank is actually needed.
    B. Compressed gas cylinders should be stored in cool areas.
    C. Properly secure tanks to avoid potential damage to the valve and/or tank.
    D. All of the above.

58. During a summer afternoon a truck was transporting 20 uncovered olive colored gas cylinders that were filled to their capacities. What potential problems, if any, could be experienced by this truck driver?
    A. Tank rupture is probably imminent.
    B. The potential of a leak to occur becomes certain.
    C. All tanks would experience an increase of internal pressure thus increasing the possibility of tank failure.
    D. No problems can be foreseen.

ANSWER QUESTIONS 59 THROUGH 66 ON THE BASIS OF THE PASSAGE BELOW

Any time firefighters are called to the scene of a suspected flammable gas leak, one of the first things they do is evacuate the affected area or building of all occupants. If it is possible to determine the source of the gas or at least the section of building affected, the windows and doors should be opened to achieve some degree of ventilation. The gas supply can be shut off by a valve inside the building or better yet terminated by closing a main valve outside the building. To verify that all gas flow has been stopped, firefighters can simply look at the gas meter provided. If the meter does not register any gas flow, it can be safely assumed that the gas source has been effectively stopped.

During any part of these procedures, an emphasis should be made to eliminate any potential ignition sources. One such precaution would be to leave light switches as they are, regardless of whether the lights are on or off. By merely flicking a switch, a small electrical spark can occur and that can be all that is required to detonate the flammable gases present. If a light source is necessary, battery operated flashlights are recommended. A more obvious precaution is to prohibit any firefighters or bystanders from smoking in the area. A struck match or dropped cigarette ash could have disastrous consequences.

Some gas leaks are easier to spot than others. If liquified gas is involved, the escaping gas manifests itself as a small vapor cloud. The reason for this is that liquified gas refrigerates the atmosphere closest to it and condenses moisture present in the air. Odorless and colorless gases are more difficult to detect. Depending on the gas involved, special instruments are needed to discern their presence.

If leaking gas has already been ignited, firefighters should not attempt to suppress the blaze until the source of gas has been cut off. If fire emerges from a relief valve on a pressurized gas cylinder, a water stream should be applied to the top of the cylinder in the hope of sufficiently cooling its contents. If this is accomplished, the relief valve will close on its own, thus ending the fire. On the other hand, if it cannot be cooled, all personnel should evacuate the immediate area. Tank ruptures eventually will occur followed by an extremely violent explosion. Gas-related fires may pose special risks to a firefighter, but if certain procedures are followed, the potential danger to personnel or structures is minimized.

59. Which of the following alternatives would be the best choice for a title for this article?
    A. Operational Procedures for Firefighters Responding to Flammable Gas Leaks or Gas Fires.
    B. The Physical Properties of Specific Flammable Gases.
    C. The Drawbacks of Smoking in the Presence of Flammable Gases.
    D. The Correct Method of Extinguishing Liquified Gas Cylinder Fires.

60. What can be said about a gas fire that sets it apart from a fire that involves ordinary combustibles such as wood, paper, or cloth?
    A. Gas fires are always preceded by the presence of a vapor cloud.
    B. Anybody in the immediate area should be evacuated.
    C. Before even attempting to extinguish a gas-related fire, the source of gas should be eliminated first.
    D. None of the above.

61. What is the reason liquified gas forms a small cloud upon its release into the air?
    A. The air is cooler than the tank's contents, thus creating a condensation effect.
    B. The escaping gas is cooler than the environment outside the tank, thus creating a condensation effect on moisture in the air.
    C. Friction.
    D. Gases have higher boiling points than liquids.

62. What was the suggested first step that firefighters should take when initially confronting a gas leak?
    A. Turn on all the lights so it is easier to view the building.
    B. Ventilate all rooms affected by the leak by opening doors and windows.
    C. Evacuate all occupants.
    D. Take a cigarette break.

63. If someone wanted to inspect a gas main suspected of leaking in a windowless basement, which of the following sources of illumination would be suggested for safety reasons?
    A. A drop light with an extension cord
    B. A self-contained chemical glow light
    C. Light a match
    D. Turn on the stairwell lights

64. How can a firefighter be sure that gas flow has been completely stopped after a valve has been turned off?
    A. Feel around gas appliances for any air currents.
    B. All hissing noises cease.
    C. Smell the air, it should be less pungent.
    D. Scrutinize the gas meter (if provided) for any incremental gas flow.

65. Which kinds of flammable gas is probably the most hazardous to personnel if a leak were to occur?
    A. Gases that are noxious to smell
    B. Gases that are heavier than air
    C. Gases that produce a vapor cloud upon being vented
    D. Colorless and odorless gases

66. When firefighters attempt to extinguish a fire coming from the safety relief valve of a liquified gas tank, a hose stream is applied to the top one-third of the tank. What is the reason for this?
    A. To eliminate oxidizing agents, thereby suffocating the fire.
    B. To guarantee that the tank's integrity is not compromised.
    C. To hopefully contract that portion of liquified gas that exists in the gaseous state within the tank. This would allow the safety relief valve to close, thus eliminating the fire.
    D. It minimizes the extent of a BLEVE (i.e., Boiling Liquid Expanding Vapor Explosion).

COMPLETE FIRE-FIGHTERS EXAM PREPARATION BOOK

ANSWER QUESTIONS 67 THROUGH 82 ON THE BASIS OF THE FLOOR PLAN BELOW

FLOOR PLAN AND CLASS LOCATION FOR A TWO STORY JUNIOR HIGH SCHOOL

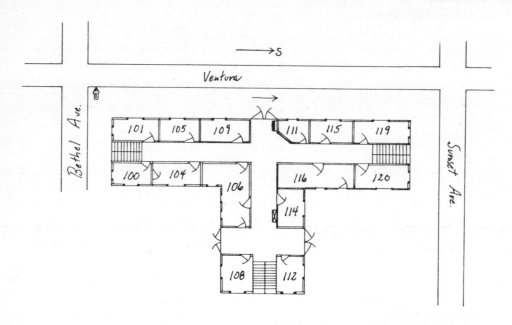

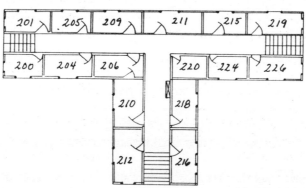

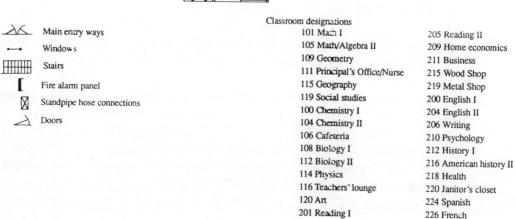

Classroom designations
101 Math I
105 Math/Algebra II
109 Geometry
111 Principal's Office/Nurse
115 Geography
119 Social studies
100 Chemistry I
104 Chemistry II
106 Cafeteria
108 Biology I
112 Biology II
114 Physics
116 Teachers' lounge
120 Art
201 Reading I

205 Reading II
209 Home economics
211 Business
215 Wood Shop
219 Metal Shop
200 English I
204 English II
206 Writing
210 Psychology
212 History I
216 American history II
218 Health
220 Janitor's closet
224 Spanish
226 French

PRACTICE EXAM 3

67. If a student was just leaving Biology II class and walking toward the principal's office, in what direction would he or she be headed?
    A. West   B. East   C. North   D. South

68. Which room presents the greatest hazard to occupants with regard to fire safety?
    A. 104   B. 100   C. 215   D. 220

69. What direction is Room 104 in relation to Room 105?
    A. West   B. East   C. North   D. South

70. If a fire originated from an inappropriate mix of chemical reagents in the Chemistry II laboratory, which rooms would be in the most immediate danger if the fire were not brought under control quickly?
    A. 100, 104, and 106
    B. 100, 101, and 104
    C. 100, 200, and 205
    D. 100, 104, 106, and 204

71. Where is the room that is designated for Chemistry I study in relation to the teachers' lounge?
    A. West   B. East   C. North   D. South

72. What is the quickest means for a student leaving writing class to get to the school nurse?
    A. Turn right on leaving the classroom and go north down the hallway to the stairs and then to the principal's office.
    B. Turn left on leaving the classroom and go north down the hallway to the stairs and then, once on the ground floor, head south to room 111 on the east side of the building.
    C. Turn right on leaving the classroom and go south down the hallway to the stairs and then, once on the ground floor, head north down the hallway to Room 116 on the east side of the building.
    D. Turn right on leaving the classroom and go to the flight of stairs in the east wing of the building and then, once on the ground floor, west to room 111 next to the main entryway.

73. Where are foreign language classes taught with respect to where business is studied?
    A. Diagonally across the hall due southwest
    B. Across the hall due east
    C. Diagonally across the hall due northeast
    D. Right next door due north

74. If there were a fire located in the southeast corner of the cafeteria, which of the following alternatives would be considered the safest and most direct for stretching hose line from the standpipe connection?
    A. Stretch the hose line east down the hallway and then enter through the east door of the cafeteria.
    B. Stretch the hose line across the hall and enter through the south door of the cafeteria.
    C. Stretch the hose line down the north corridor and enter through the west door of the cafeteria.
    D. Stretch the hose line down the north corridor to the outside of the building and send a hose stream through the windows on the north side of the cafeteria.

75. If firefighters responded to a fire of unknown origin in the junior high school, which approach below would probably be the most efficient in terms of time spent trying to locate the fire? (Note: the fire department is located four blocks northeast of the school.)
    A. Drive a fire apparatus west on Bethel Avenue and park directly in front of the school's south wing; enter the south corridor and head to the fire-alarm panel on the south wall of the principal's office.
    B. Drive the fire apparatus south on Ventura and west on Sunset Avenue and then park directly in front of the south wing of the school; enter the south corridor and go directly to the fire alarm panel located on the north wall of room 111.
    C. Drive the apparatus west on Bethel Avenue and park directly north of the north entrance to the west wing of the school; enter the north corridor and then proceed down the east corridor directly to the fire panel located to the right of the east main entryway.
    D. Drive the apparatus south on Ventura and park directly in front of the east main entryway to the school; enter the east main entryway and immediately look left to the fire alarm panel.

76. Which of the following statements is true?
    A. Psychology class is conducted directly below the cafeteria.
    B. The two rooms that are the farthest apart from one another are rooms 200 and 109.
    C. A fire restricted to room 201 would pose less of a hazard to people than a fire in room 206.
    D. Business class is conducted in a room that is directly above where geometry is taught.

77. Without regard to a fire location, what would be the best route for health students to take when evacuating the building?
    A. Walk down the east corridor and then go south to the stairwell. Once on the ground floor, head north just past the principal's office and leave through the east main entryway.
    B. Get to the west wing stairwell and leave through the south main entryway.
    C. Get to the north wing stairwell and leave through the north main entryway.
    D. Get to the south wing stairwell and leave through the south main entryway.

78. Referring to question 77, what would be the safest means to evacuate if smoke were rising up the west wing stairwell?
    A. Get to the south wing stairwell and leave the premises through the east main entryway.
    B. Crawl out the windows on the south side of the room.
    C. Head cautiously down the west wing stairwell and then leave through either the north or south main entryway, depending on which is the safest.
    D. Get to the north wing stairwell and exit the premises via the north main entryway.

79. What room in the school has the largest number of possible fire exits?
    A. Room 106
    B. Room 209
    C. The room where Biology I is studied
    D. The room where social studies is taught

80. In a locational sense, which of the choices below reflect the greatest distance between two rooms?
    A. Woodshop and writing classrooms
    B. Math I and Reading II classrooms
    C. Physics and American History II classrooms
    D. English I and social studies classrooms

81. If the physical education building were a separate structure approximately 100 yards southwest of the junior high, what would the name of the street that runs closest to it be?
    A. Bethel Avenue
    B. Ventura
    C. Sunset Avenue
    D. It cannot be determined by the diagram provided.

82. Where is geometry studied in relation to where home economics is studied?
    A. East
    B. Southeast
    C. One floor directly above
    D. One floor directly below

83. The illustration shown below is one kind of water pump used by firefighters. Judging by the impeller's design, what kind of force is chiefly employed to convey water through the outlet port? Water enters from a lateral conduit at the base of diffusion vanes and ultimately leaves through the outlet port.

    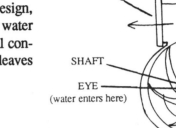

    A. Gravitational force
    B. Centrifugal force
    C. Osmotic pressure
    D. Negative pressure because of a vacuum

84. What is the tool illustrated below principally used to do?

    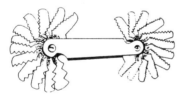

    A. Cut small holes in extremely thin wood.
    B. Determine the pitch of a machined thread.
    C. Etch round metal stock.
    D. Adjust spark plug gaps.

85. One kind of fire hose coupling has recessed lugs spaced evenly on its exterior collar. If a firefighter needed to break a coupling of this nature, which of the wrenches shown below would probably be used?

    A.
    B.
    C.
    D.

86. The illustration to the right is called a Siamese hose adaptor. When considering that water enters through the female connective end and exits the male connective end, this device has what kind of an effect on the resulting water flow?

   A. It reduces the potential discharge by at least 50%.
   B. It splits the water flow, which increases frictional loss.
   C. It triples the potential discharge.
   D. It combines the potential discharge of two separate hose lines into a master fire stream.

87. Which of the power tools below is best used to put a smooth finish on hardwood?

   A.   B.   C.    D.

88. Which of the tools illustrated below is best used to sufficiently heat ½-inch copper conduit to take solder when forming a joint?

   A.    B.    C.    D.

89. What one tool below is not used in the automotive trade?

   A.    B. 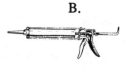   C.    D.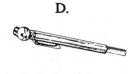

90. If an individual wanted to dig a hole approximately two feet deep in the soil and was aware that the soil was composed of a mixture of very hard clay and gravel, the best tool suited for the job would be:

   A.    B.   C. 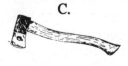   D.

PRACTICE EXAM 3

ANSWER QUESTIONS 91 THROUGH 93 ON THE BASIS OF THE PUMP DIAGRAM BELOW

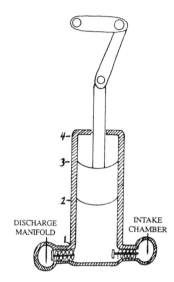

91. Considering how the valves are designed on this pump, what will happen when the driveshaft turns ¾ of a revolution counterclockwise from its present position?
    A. The cylindrical volume of water purged through the discharge manifold is represented by the amount of water between points 1 and 2.
    B. The cylindrical volume of water purged through the discharge manifold is represented by the amount of water between points 1 and 3.
    C. The cylindrical volume of water purged through the discharge manifold is represented by the amount of water between points 1 and 4.
    D. The cylindrical volume of water purged through the discharge manifold is represented by the amount of water between points 2 and 4.

92. By observing the pump action described in the previous question, we can tell at what point a vacuum ceases to exist in drawing water through the intake chamber and positive displacement of water out the discharge manifold begin?
    A. Once the driveshaft has turned counterclockwise ¼ of a revolution.
    B. Once the driveshaft has made a complete revolution.
    C. Once the driveshaft has turned counterclockwise ½ of a revolution.
    D. Once the driveshaft has turned counterclockwise ¾ of a revolution.

93. Which of the following statements is true with regard to this pump's mechanical features?
    A. As the driveshaft approaches its top stroke, the discharge manifold valve remains open.
    B. As the driveshaft approaches its down stroke, the intake valve remains open.
    C. As the driveshaft approaches its top stroke, the intake valve remains closed.
    D. When the driveshaft is at maximum down stroke, water is still trapped between the two valves.

94. If a fire hose coupling were partially crushed by a fire apparatus running over it during a training exercise, which of the following statements is true. (Assume the coupling seal is maintained while the hose is charged.)
    A. In terms of volume, the same number of gallons of water per minute will pass through the hose line leading up to the coupling as the number of gallons of water per minute that will pass through the bent coupling itself.
    B. The flow velocity of water will be greater once it gets through the bent coupling.
    C. In terms of gallons per minute, a greater number of gallons flow through the hose line leading up to the coupling than the number of gallons that pass through the coupling.
    D. The flow velocity of water will be greater in the hose line leading up to the coupling than passing through the coupling itself.

ANSWER QUESTIONS 95 THROUGH 97 ON THE BASIS OF THE 2½ INCH WYE HOSE ADAPTOR ILLUSTRATED.

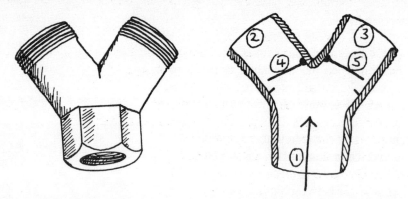

The appliance shown is called a *wye* and it is essentially used to split a master hose stream into two smaller lines. This kind of appliance can be manually gated; however, for our purposes the interior cutaway view demonstrates two check valves. The arrow indicates the direction of water flow.

95. If the two smaller hand lines were of equal length, but line 2 was carried up approximately three flights of stairs while the other line was used to attack a fire at ground level, which of the following would occur?
    A. Both valves remain in the open position and supply both hand lines an equal amount of water.
    B. Valve 5 closes permitting water to flow only through hose 2.
    C. A backpressure created in hose 2 will close valve 4 if it exceeds pump pressure.
    D. Valves 4 and 5 will close simultaneously if pressure loss from gravity is enough to overcome pump pressure in either line.

96. Assuming a firefighter mistook this wye appliance for a Siamese hose adaptor and through various couplings managed to hook up Lines 2 and 3 to two different pumpers, what would be the end result?
    A. The combined pressures of hose lines 2 and 3 would serve to create a singular master stream out of point 1.
    B. If the pump pressure in line 3 exceeded that of line 2, check valve 4 would automatically close.
    C. If the pump pressure in line 2 greatly exceeded that in line 3, valve 5 would automatically close.
    D. Both valves would prevent any water flow, thus creating a static pressure in both lines.

97. If the two smaller hand lines were of equal length, and line 3 was battling a ground fire at the same center line (i.e., level), but experienced diminished hose pressure, what can be assumed if the pump pressure remains constant?
    A. Line 3 had backpressure that needed to be accounted for.
    B. Line 2 must have been dragged to a level below the pump's centerline and the pressure gain acquired diverted most, if not all, of the water to that line.
    C. Line 2 had been elevated to a point above the pump's centerline, so the backpressure that resulted closed valve 5.
    D. Valves 4 and 5 must have malfunctioned and remain in a partially closed position.

98. A 10-foot section of an 8-inch cast iron main was suspended by three metal straps from a basement ceiling. Knowing there was a strap supporting both ends and the center strap was situated closer to one end than the other, which of the statements given below would be true?
    A. The strap on the end that is furthest away from the center supporting strap would bear the greatest load.
    B. All three straps would bear equal load.
    C. The end strap closest to the center support strap would bear the greatest load.
    D. The center supporting strap would bear the greatest strain.

99. Which of the situations given below is more conducive for evaporation to take place?
    A. A narrow, but tall cylinder filled with water without a lid on.
    B. Water spilled over a large area of pavement.
    C. Water in an open container such as a bucket.
    D. Water in a closed container.

100. In the illustration shown, if the drive gear is the larger gear, what overall effect does the smaller gear have on the drivetrain?

    A. There is a proportional increase in speed.
    B. There is a proportional decrease in both torque and speed.
    C. There is a proportional increase in speed but a proportional decrease in torque.
    D. All factors considered, it is a 1:1 gear ratio and it does not have any impact on speed or torque.

ANSWER SHEET TO EXAM 3

1. (A) (B) (C) (D)
2. (A) (B) (C) (D)
3. (A) (B) (C) (D)
4. (A) (B) (C) (D)
5. (A) (B) (C) (D)
6. (A) (B) (C) (D)
7. (A) (B) (C) (D)
8. (A) (B) (C) (D)
9. (A) (B) (C) (D)
10. (A) (B) (C) (D)
11. (A) (B) (C) (D)
12. (A) (B) (C) (D)
13. (A) (B) (C) (D)
14. (A) (B) (C) (D)
15. (A) (B) (C) (D)
16. (A) (B) (C) (D)
17. (A) (B) (C) (D)
18. (A) (B) (C) (D)
19. (A) (B) (C) (D)
20. (A) (B) (C) (D)
21. (A) (B) (C) (D)
22. (A) (B) (C) (D)
23. (A) (B) (C) (D)
24. (A) (B) (C) (D)
25. (A) (B) (C) (D)
26. (A) (B) (C) (D)
27. (A) (B) (C) (D)
28. (A) (B) (C) (D)
29. (A) (B) (C) (D)
30. (A) (B) (C) (D)
31. (A) (B) (C) (D)
32. (A) (B) (C) (D)
33. (A) (B) (C) (D)
34. (A) (B) (C) (D)
35. (A) (B) (C) (D)
36. (A) (B) (C) (D)
37. (A) (B) (C) (D)
38. (A) (B) (C) (D)
39. (A) (B) (C) (D)
40. (A) (B) (C) (D)
41. (A) (B) (C) (D)
42. (A) (B) (C) (D)
43. (A) (B) (C) (D)
44. (A) (B) (C) (D)
45. (A) (B) (C) (D)
46. (A) (B) (C) (D)
47. (A) (B) (C) (D)
48. (A) (B) (C) (D)
49. (A) (B) (C) (D)
50. (A) (B) (C) (D)
51. (A) (B) (C) (D)
52. (A) (B) (C) (D)
53. (A) (B) (C) (D)
54. (A) (B) (C) (D)
55. (A) (B) (C) (D)
56. (A) (B) (C) (D)
57. (A) (B) (C) (D)
58. (A) (B) (C) (D)
59. (A) (B) (C) (D)
60. (A) (B) (C) (D)
61. (A) (B) (C) (D)
62. (A) (B) (C) (D)
63. (A) (B) (C) (D)
64. (A) (B) (C) (D)
65. (A) (B) (C) (D)
66. (A) (B) (C) (D)
67. (A) (B) (C) (D)
68. (A) (B) (C) (D)
69. (A) (B) (C) (D)
70. (A) (B) (C) (D)
71. (A) (B) (C) (D)
72. (A) (B) (C) (D)
73. (A) (B) (C) (D)
74. (A) (B) (C) (D)
75. (A) (B) (C) (D)
76. (A) (B) (C) (D)
77. (A) (B) (C) (D)
78. (A) (B) (C) (D)
79. (A) (B) (C) (D)
80. (A) (B) (C) (D)
81. (A) (B) (C) (D)
82. (A) (B) (C) (D)
83. (A) (B) (C) (D)
84. (A) (B) (C) (D)
85. (A) (B) (C) (D)
86. (A) (B) (C) (D)
87. (A) (B) (C) (D)
88. (A) (B) (C) (D)
89. (A) (B) (C) (D)
90. (A) (B) (C) (D)
91. (A) (B) (C) (D)
92. (A) (B) (C) (D)
93. (A) (B) (C) (D)
94. (A) (B) (C) (D)
95. (A) (B) (C) (D)
96. (A) (B) (C) (D)
97. (A) (B) (C) (D)
98. (A) (B) (C) (D)
99. (A) (B) (C) (D)
100. (A) (B) (C) (D)

ANSWERS TO PRACTICE EXAM 3

Refer to the pump-panel illustration for any clarification on questions 1 through 17.

1. D	7. D	13. B
2. B	8. B	14. D
3. A	9. A	15. D
4. C	10. C	16. B
5. C	11. A	17. C
6. D	12. C	

18. **C.** Since heat from any combustion is known to rise, a basic rule of thumb is to create a ventilation port directly above the fire. Choices A and B would make the fire worse by either fanning the flames more or drawing the fire to a wall that might not otherwise have been involved. Choice D is incorrect because the opening made in the adjoining room would not effect proper ventilation for the fire in the bedroom.

19. **B.** A skull and crossbones is on many household cleaning products and indicates that the contents are poisonous to people; definitely not an inert substance as implied by choice A. Choices C and D make dangerous assumptions. Some substances have only to be breathed to be toxic. Many chemical spills can evolve lethal vapor clouds or at least present high enough concentrations to make someone extremely sick. When it comes to cleaning up a toxic spill, there is nothing routine about it. The chemical(s) involved need to be fully identified and understood before any cleanup is attempted. Dilution with a hose stream without knowledge is extremely dangerous.

20. **D.** The line tied to the underside of the ventilation fan is known as a tag line. When this is kept semitight at a point 15 to 20 feet from the base of the building, it will help prevent the equipment from striking the side of the building. The remaining options are false.

21. **C.** If a label on a fire extinguisher reveals a red slash through symbols for "flammable liquids" and "electrical equipment," the unit could be applied only to ordinary combustibles such as wood, paper, or cloth. This is not a very versatile fire extinguisher.

22. **D.** Any fire extinguisher that bears the symbol for "electrical equipment" with a slash through it indicates that the agent contained within that extinguisher can potentially conduct electrical current. In other words, it presents an electrical hazard to the user if the agent comes into contact with an energized line. Choices A and B are false. Choice C is absolutely ludicrous because anything that is alcohol based is considered flammable, not an extinguishing agent.

23. **D.** Only choice D would be the correct procedure for anchoring hose line as an apparatus moves forward. The other options are conducive to injury. Safety is always a priority regardless of what kind of a procedure is being performed.

24. **A.** The V-type and lean-to collapse can create voids on the floor beneath, thus sheltering victims from falling debris. Shaft construction or tunneling are two means of extricating victims from

such voids. A pancake collapse, however, leaves little chance for occupant survival. Roofs have been known to completely collapse to basement level.

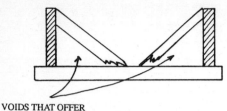
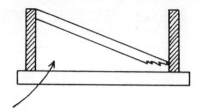

VOIDS THAT OFFER
POTENTIAL SHELTER

25. *B.* Choices A, C, and D are clues that can identify hot spots within a wall's interior. Swelling or contraction of a partition does not indicate a hot spot.

26. *A.* Since 26 days of work were lost due to eye injuries over a 10-year period, an average of 2.6 days were lost per year (i.e., 26 ÷ 10 = 2.6).

27. *C.* Since 37 days of work were lost due to smoke inhalation over a 10-year period, an average of 3.7 days were lost per year (i.e., 37 ÷ 10 = 3.7).

28. *C.* Wet-barrel fire hydrants are at a disadvantage when it comes to freezing weather. If ice forms within the hydrant, not only could water flow be impeded, but there is the strong possibility that the hydrant will rupture. Dry-barrel fire hydrants, on the other hand, circumvent this problem. Choices A, B, and D are false.

29. *D.* Choices A and C are incorrect because of the presence of check valves. Choice B is incorrect because it was stated in the question that the manual valve identified as H was in the closed position. Under the circumstances given, water would only flow out both K and L outlet ports. One other thing that should have helped in determining the answer to this question is that choices A and C are referred to as outlet ports. This is an incorrect assumption on the basis of the check valve position.

30. *D.* The key valve here is the manual valve indicated by the letter F. It is in the closed position, water under pressure entering through inlet port B would have no place to flow because of check valve positioning in the other ports (i.e., A, C, and E).

31. *A.* The water would leave the outlet port identified by the letter I. Choices B and C are incorrect because manual valve J is in the closed position. Water cannot enter the inlet port identified by the letter E because of the check valve.

32. *D.* Firefighting is a continual learning process. New recruits go through extensive training in the fire academy and are then placed in a subsequent probationary period, but that does not necessarily guarantee that everything is going to be done correctly all the time. Considering the way this question is worded, the infractions involved only the misuse of some equipment and did not imply anything life threatening. Choice D would be the appropriate manner to handle this kind of situation. Discipline would be necessary only if the same infraction were to continue.

33. *B.* Only those factors in choice B have direct bearing on fire progression. Number 3 is important in starting a fire, but it does not impact what follows in terms of progression. Numbers 5 and 8 are insignificant in regard to fire progression. Number 7, as it is stated, means very little.

34. *C.* Since heat rises, any time a building has an unprotected vertical artery such as an open staircase, fire will spread significantly faster as compared to lateral extension when fire is essen-

tially confined to one floor (i.e., spread room to room versus floor to floor). Enclosed stairwells and elevator shafts, on the other hand, can serve as fire barriers, provided the doors remain closed or sealed.

35. **D.** This choice is the correlation that can be drawn from the passage. The other choices contradict what is stated.

36. **B.** If an uncharged hose is accidently laid beneath a door, once the line becomes charged the door can effectively serve as a clamp and restrict water flow to the nozzle. Doors can normally be removed by taking them off their hinges, but the best way to prevent this is to avoid laying hose under doors in the first place. Choices A and D are remote possibilities and choice C can be true indirectly; if firefighters have to forcibly remove the door.

37. **B.** $$\frac{4}{5} \times \frac{5}{8} = \frac{20}{40} \;;\; \frac{20}{40} = \frac{1}{2} \text{ when reduced.}$$

    Therefore, a ½-inch tap would be required.

38. **C.** Simply divide 62⅖ pounds/cubic foot into 2402½ pounds to determine how many cubic feet of water are in the tank.

    $$2{,}402\tfrac{1}{2} \div 62\tfrac{2}{5} = X$$

    $$\frac{4805}{2} \div \frac{312}{5} = X$$

    $$\frac{4805}{2} \times \frac{5}{312} = X \quad X = \frac{24{,}025}{624} = 38.5 \text{ cubic feet}$$

39. **A.** 16,000 pounds divided by 8³⁵⁄₁₀₀ pounds = X gallons of water.

    $$16{,}000 \div \frac{835}{100} = X \quad 16{,}000 \times \frac{100}{835} = X$$

    $X = 1{,}916.17$ gallons of water.

40. **A.** A hexagonal nut has six sides. We can think of the nut as being a six-sided polygon. Since each side is equal in length, we can simply multiply the length of one side by 6 to determine perimeter length. Therefore,

    $6 \times \tfrac{2}{3} = \tfrac{12}{3} = 4$ centimeters

41. **A.** The difference is equal to the factory specs minus the micrometer reading.

    $1.750 - 1.732 = .018$ inches

42. **D.** What first must be determined is how much dry chemical agent was used to extinguish the fire. 8½ lbs - 3¼ lbs = X
    The LCD is 4. 8²⁄₄ - 3¼ = X lbs.

$X = 5¼$ lbs.

Now, to determine what percentage 5¼ lbs. represents, simply divide 5¼ by its capacity, 8½ lbs., x 100. Therefore, $5¼ ÷ 8½ × 100 = X\%$

$$\frac{21}{4} \times \frac{2}{17} \times 100 = \frac{4200}{68} = 61.76\%$$

43. **C.** Fire District 18 had 39 fewer fires in 1989 than in 1988. To figure the percentage this difference corresponds to, we need to divide 39 by 252 and multiply it by 100. Therefore,

$$\frac{39}{252} = .15476$$

.15476 x 100 = 15.476%, or 15.5% when rounded.

44. **A.** Fire District 12 had four more cases of arson in 1989 than in 1988.

$$\frac{4}{14} \times 100 = 28.57\% \text{ increase.}$$

Fire District 18 had 15 more cases of arson in 1989 than in 1988.

$$\frac{15}{32} \times 100 = 46.87\% \text{ increase}$$

Fire District 22 had two more cases of arson in 1989 than in 1988.

$$\frac{2}{1} \times 100 = 200\% \text{ increase}$$

45. **D.** Since 14 fatalities resulted from fire in Fire District 12 during 1988, we can figure the number of fatalities in 1992 by setting up the problem as shown. (Fatalities in 1992 = X).

$$\frac{X - \text{Fatalities in 1988}}{\text{Fatalities in 1988}} = 4.5$$

Therefore,

$$\frac{X - 14}{14} = 4.5$$

$X - 14 = 63$; $X = 63 + 14$ or 77 fatalities in 1992.

46. **D.** The hose nozzle weighs 4.75 pounds and is composed of 67% copper. 4.75 x .67 = 3.1825 pounds. This number represents the weight of copper in pounds present in the alloy. Since there are 16 ounces per pound, we can multiply 3.1825 by 16, which will give us the amount of copper present. X = 50.92 ounces.

47. C. Ratios should be reduced to their lowest terms.

Therefore

$$58 : 8 = \frac{58}{8} : 1 \text{ or } 7.25 : 1.$$

48. D. A cubic decimeter is 1/1000 of a cubic meter. Therefore,

$$\frac{675 \text{ MG}}{1 \text{ DM}^3} \times \frac{1000 \text{ DM}^3}{1 \text{ M}^3} = 675{,}000$$

micrograms of asbestos would be present in a cubic meter. It should be noted here that liters are normally used to measure liquid volumes but they can be used to figure other volumes as well.

49. A. Using the Pythagorean theorem, we can figure the length of guy wire required (which represents the hypotenuse of a right triangle) by squaring the height of the collar plus squaring the base.

$$23^2 + 15^2 = X^2$$
$$529 + 225 = X^2$$
$$754 = X^2 \quad X = \sqrt{754} = 27.46 \text{ feet.}$$

The question also stated that an extra 8 inches would be required to fasten the ends of each wire. Since we are dealing in units of feet; 8 inches = 8/12 = .67 feet. Therefore, the total length of one wire is equal to 27.46 feet + .67 feet or 28.13 feet. Four such lengths of wire are required to support the antenna, so 4 x 28.13 feet = 112.52 feet of guy wire is needed.

50. D. It is best to draw a picture of what is described in the question to avoid any confusion. A room with the dimensions 25 feet x 27 feet x 9.6 feet would look like the example shown to the right. Since drywalling is not utilized for flooring, we are essentially looking at a rectangular geometric shape minus the bottom. The ceiling accounts for 25' x 27' or 675 square feet. The two walls measuring 27' x 9.6' account for 518.4 square feet and the other two walls measuring 25' x 9.6' account for 480 square feet. Total square feet of drywalling required for this room then is 675 sq. ft. + 518.4 sq. ft. + 480 sq. ft. or 1,673.4 sq. ft. However, this number does not take into account the door and windows. The square footage of doors and windows should be subtracted from this total to determine the actual amount of drywall needed.

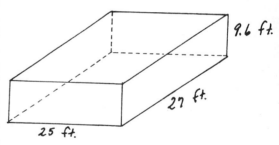

$$1{,}673.4 \text{ sq. ft.} - 48.5 \text{ sq. ft.} = 1{,}624.9 \text{ sq. ft.}$$

Choice C does not account for the square footage of drywall required for the ceiling.

51. C. The volume of a sphere is equal to (4/3) x π x R³. The radius of this sphere equals 72.75 cm x .5 or 36.375 cm.

(4/3) x 3.1416 x 36.375³ = X cubic centimeters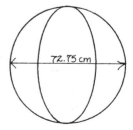
(4/3) x 3.1416 x 48,129.239 = X
X = 201,603.76 cubic centimeters

52. *B.* This choice best covers the article as a whole. The potential hazards were elaborated on in the first part of the article, followed by a short overview of how compressed and liquified gases can be safely handled. Choices A and C are essentially one and the same. They both touch on individual topics discussed in the reading but fail to convey the entire meaning of the article. Choice D is probably the next best selection. As it reads, it would indicate that the article was more slanted toward employee safety than anything else. However, the article placed an equal emphasis on how various situations may affect a tank containing liquified or compressed gases.

53. *C.* The four were (a) physical damage to the tank, (b) combustion of the tank's contents, (c) heat-caused gas expansion, and (d) heat-induced metal fatigue of the cylinder wall.

54. *D.* One sentence in the article mentioned that if a tank is filled to its capacity and then subjected to a sufficient amount of heat, an excess pressure can build, thus exceeding the tank's limitations. The other options can have an affect on tank integrity, but none were discussed in the reading.

55. *A.* Static electricity was said to play a small or insignificant role when considering any closed system. Choices B, C, and D are all true statements according to the passage.

56. *A.* Safety relief valves are installed principally to bleed off excess pressure that accumulates within a gas cylinder. Choices B and D are false. Choice C is incorrect, too, because the key word here is *any*. It may circumvent tank rupture under certain conditions; however, if a tank is subjected to enough heat, the presence of a safety valve is a moot point.

57. *D.* All three alternatives were suggested safety guidelines for handling compressed gas cylinders.

58. *C.* We can assume by how the question is worded that it is a warm, if not hot, summer afternoon. Since dark colors absorb heat instead of reflecting it, the fact that the cylinders in question are dark colored raises the specter of heat absorption. As the contents are heated, a proportional increase in pressure results, enhancing the possibility of tank rupture. This is especially true considering that the tanks being transported are at full capacity to begin with. Tank rupture is not imminent, as indicated in choice A, but the possibility becomes more significant. Choice B is essentially the same thing as a tank failure. It is by no means a certainty; however, it can become significant as well.

59. *A.* This choice encompasses what is explained in the entire article. Choices B, C, and D are individually touched on but fail to summarize the passage.

60. *C.* As mentioned in the reading, statement C is correct. Choice A implies that all gas fires are preceded by the presence of a vapor cloud. This is a correct assumption with regard to liquified gases; however it is not true for all other gases. Choice B is a similarity, not a difference in operational procedures for firefighters. The question asks what difference exists between a gas fire and another fire involving ordinary combustibles, not what factors are common to both.

61. *B.* Vented liquified gas has a refrigerating effect on the air around it. The moisture in the air condenses to some extent, causing a vapor cloud to materialize.

62. *C.* It is suggested to clear the area of all occupants or bystanders and then proceed with ventilation. Choices A and D are wrong because each may be responsible for igniting the gas present.

63. *B.* As mentioned in the article, battery-operated flashlights are recommended. However, this option was not given as a possible answer. Realistically, the gas company or fire department should be notified if a gas leak is suspected. If one of the four answers had to be selected on the basis of safety, the chemical glow light would stand out. What should have been a clue is

the fact that it is self-contained, therefore eliminating the possibility of an open spark as a drop light or light switch may present. Chemical glow lights are tubes containing two different chemicals. When mixed, they offer sufficient illumination but produce a minimum of heat. Consequently, they would be safe in such a situation.

64. **D.** Providing a gas meter is present, it can easily be determined if any gas flow were to continue. It may register a small reading but nonetheless make it obvious whether or not a shut off valve is effectively closed. The other choices are unreliable options and are considered unsafe.

65. **D.** Colorless and odorless gases are especially dangerous because they are impossible to detect without the aid of specialized devices that can discern their presence. Most other gases when released provide either a visual clue or have a characteristic smell that can alert people to their presence. Appropriate measures can then be taken.

66. **C.** A hose stream applied to the top third of a liquid gas cylinder has a cooling effect on the tank itself. Its purpose is twofold. First, it cools that portion of the gas that exists as a liquid. Second, and more important, it cools the liquid gas that already exists in the gaseous state. These gases tend to accumulate in the topmost portion of a cylinder. As these gases are cooled, a contraction results, with a corresponding decrease of pressure. At some safe point, the safety relief valve closes, eliminating the fire altogether. Using a hose stream is not sufficient to completely eliminate all oxidizers necessary for a gas fire to be suppressed, as suggested in choice A. Gas fires are particularly stubborn. Choice B is wrong because you cannot guarantee that a tank's integrity has not been compromised by simply applying a hose stream to it. Several factors need to be taken into consideration before any such conclusion can be drawn. Choice D is incorrect because there is no such thing as minimizing a BLEVE. Once the conditions are right for this to take place, an explosion of the same intensity will occur.

67. **B.** A student walking in that direction would be headed east.

68. **D.** The concept stressed here is fire safety. Granted, chemistry and wood shop classes present a spectrum of various fire hazards; however, the number of exits each room provides is the key element. According to the floor plan, room 220, which is the janitor's office and supply room, falls short in this area. It has only one possible exit. There isn't even an exterior window to provide an emergency secondary exit if a fire were to block the doorway to the hall.

69. **A.** Room 104 is west of room 105.

70. **D.** Chemistry II lab is conducted in room 104. If a fire were to originate there, the rooms sharing the same partitions with this one would be in danger. The room which is directly above the Chemistry II lab (room 204) would be in worse danger due to the vertical extension characteristic of fire. Therefore, choice D includes the rooms that would be immediately affected. The other alternatives given either did not account for an upstairs room being affected or mentioned a room on the ground floor not immediately proximate to the fire.

71. **C.** Chemistry II is taught in room 100. This particular room is located north of the teachers' lounge.

72. **B.** Turning left upon leaving the writing class would in fact mean heading in a northerly direction. Once the student descends the flight of stairs to the ground floor, the student would need to walk in a southerly direction to get to the nurse located in room 111, which is situated on the east side of the hallway immediately after the main entry. Choice A is incorrect because making a right turn after leaving writing class the correct orientation for the student would be southerly instead of north. Choice C is incorrect because the nurse is not located in room 116 according to the room designation key in the illustration. Choice D is wrong, too, because there is no east wing to this building, according to the floor plan.

73. A. Foreign language courses are taught in rooms 224 and 226 and business is taught in room 211. According to the floor plan, the foreign language classes are diagonally across the hall from room 211 due southwest. If the question had been worded, "where is business class taught with respect to foreign language studies?" choice C would have been the correct one. Choice B is incorrect because metal and wood shops are east of where foreign language is studied. Choice D is wrong because it describes the location of room 220 (janitor's office and supply room).

74. C. The west door of the cafeteria would offer the best approach because if the fire is known to be confined for the moment in the southeast corner of the cafeteria, this direction affords firefighters a straight line of attack. Firefighters attempting choices A or B as alternative approaches may find these approaches directly blocked by the fire, or the door may encumber their ability to directly attack the fire. Choice D may seem correct, but it is normally safer and more efficient for firefighters to enter doorways and attack a fire than to use a hose stream directed through a window.

75. D. Choice D gives the most direct approach that could be taken. Choice A is incorrect because it is the north wing of the school that faces Bethel Avenue. Choices B and C are both wrong as well because they require approaching the fire alarm panel in an indirect manner, which wastes precious time.

76. C. This is the only true statement because room 201 shares a partition only with room 205. It also helps that the room is on the top floor instead of the ground floor because this will limit the fire's ability to vertically extend itself to other quarters of the school. Room 206, on the other hand, shares partitions with two other rooms (204 and 210). This increases the likelihood of a fire laterally extending itself to other areas. The remainder of the choices are false on their own merit.

77. B. Choice B is the shortest and quickest means to evacuate the building from room 218. The other alternatives given are means of escape; however, they are indirect ways to evacuate the building and therefore would require extra time and pose unnecessary hazard to the people involved.

78. A. Choice A would be the safest means to evacuate the building under the circumstances. It lessens the chance of being exposed to either fire or smoke. Choices C and D are incorrect because it unnecessarily imperils school children by fire and smoke hazard. Judging from where the smoke is coming from initially, it is best to assume that there is a fire somewhere on the ground floor in the west wing. According to the floor plan, this potentially blocks the north and south main entryways. Choice B should be used only as a last resort for escaping fire.

79. A. Room 106 (the cafeteria). It has six exits altogether, three windows and three doors. Choice B, the home economics classroom, has only three exits altogether (two windows and one door). Choice C, room 108, has the same number and kind of exits as choice B. Choice D, room 119, has the same number and kind of exits as choice B or C.

80. D. Choice D is the correct answer. Not only do you have to descend the north wing stairwell, but it is necessary to walk the entire length of the building to get to room 119. The other choices are significantly shorter by comparison.

81. C. Sunset Avenue would be the closest street to the physical education center if it were located 100 yards southwest of the main building. The other choices are incorrect.

82. D. The geometry classroom is directly beneath home economics which is upstairs. Choice C would have been the correct choice if the question had been worded, "where is home economics studied in relation to where geometry is studied?"

ANSWERS TO PRACTICE EXAM 3

83. **B.** The rotation direction, diffusion vane curvature, and flow of water should have been clue enough that a pump of this design operates on the principle of centrifugal force. As water enters the eye, it is thrown or forced to the exterior wall of the pump. This creates the necessary force for water to be conveyed through the outlet port. The speed at which the impeller turns directly determines the amount of water discharged. Choice D is used more by positive displacement pumps that incorporate negative vacuum for water intake.

84. **B.** The tool illustrated is a screw pitch gauge. It is principally used as suggested in Choice B.

85. **C.** Spanner wrenches are designed with this very purpose in mind. Choice A is an Allen wrench used to tighten or loosen machined setscrews. Choice B is a torque wrench used to apply measured force to nuts or bolts. Choice D is a tap wrench used in conjunction with a tap when cutting internal screw threads.

86. **D.** To further clarify the question, the male end of a hose coupling is the end with exposed threads. When water from two separate hose lines is combined by a Siamese adaptor, the discharges of both lines are combined into a master fire stream. A hose coupling with the opposite kind of threading is termed a wye adaptor, it serves to divide a master hose stream into two separate, smaller hand lines. The other choices provided are incorrect.

87. **A.** Choice A is an illustration of an orbital sander, which best serves this purpose. Choice B is a belt sander, which is used for coarser work that must be accomplished in an expedient manner. Choice C is not a power tool. Choice D is a disk sander that performs similarly to an orbital sander providing that fine sandpaper is used. The orbital sander, however, can attain the smoothest finish of all the alternatives given.

88. **C.** A propane torch is the most appropriate tool to use for such a job. Choice A fuses metal to metal by utilizing electrical current and a welding rod; solder is not involved. Choice B is a soldering gun, however it is more appropriately applied to smaller scale projects such as wire or contact terminal connections. Choice D is an acetylene torch, and it can work if used with some finesse. Its primary use is for cutting thick metal workpieces.

89. **B.** Choice B is an illustration of a caulking gun, which is used principally to apply sealants. The other choices, a hydrometer, grease gun, and tire pressure gauge in that order, have automotive applications discussed earlier in this book.

90. **B.** Choice B is principally used to break up hard material like the soil described in the question. Choice A would be more appropriate in loamier (i.e., looser) soils. Both choices C and D are used principally to hew wood.

91. **B.** The diagrams below illustrate the sequence of pump action that takes place and the volume of water vacuumed and then purged through the discharge manifold.

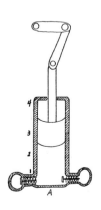
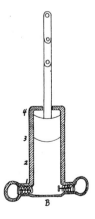
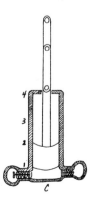

92. A. When the pump is at its top stroke as shown in the middle figure above, negative pressure/vacuum ceases to exist and positive displacement of water out of the discharge manifold begins. Therefore, for the pump to reach its top stroke in a counterclockwise fashion from the starting position given, it requires only a ¼ revolution of the driveshaft.

93. D. The water trapped within the space described is termed residual volume. Insufficient positive pressure exists to force the discharge manifold valve to remain open. For the short period of time that exists before a vacuum is created by the driveshaft upstroke, a small residual amount of water remains trapped between the two valves. The other choices given are false on their own merit.

94. A. Since the bent coupling serves as a restriction point in the hose lay, only statement A is correct. The water volume permitted to squeeze past this restriction in the hose line determines the volume of water that flows through the hose leading up to the coupling. Choice B is incorrect because flow velocity leaving the coupling would be slowed significantly. If choice C was correct, a hose rupture would become imminent somewhere between the bent coupling and the pressure source. Choice D is incorrect for basically the same reason choice C was.

95. C. As a hose is lifted above the centerline of a pump, there is a corresponding loss of pressure due to gravity. In fact, pressure loss is approximately .5 pounds per square inch (psi) per foot. If the hose length is lifted to a point high enough, a back pressure is created that may well exceed the pump's capacity. In this case, valve 4 would close, thus diverting 100% of the flow to the ground line. The other choices given are incorrect as stated.

96. D. If water intake were reversed and water came from the two smaller lines, the valves, as they are designed, would close against their backstops and thus completely stop the water flow. The remainder of the choices are incorrect with respect to the kind of hose connections described in the question.

97. B. As a hose is carried downhill from the centerline of a pump, there is a corresponding pressure gain due to gravity. Essentially, the opposite occurs to what was explained in answer 95. The resulting pressure gain can actually cause a suction effect on the other handline causing valve 5 to close.

98. A. The question depicts what is diagrammed to the right. In this case, strap 1 would be subjected to the greater load factor of the three strap supports shown. Choice B would have been correct if the strap supports were evenly spaced.

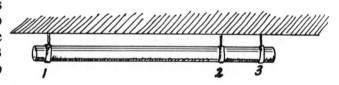

99. B. Evaporation is defined as the escape of a liquid into the air as a vapor. This occurrence is directly proportional to the surface area of a liquid exposed to the air. Therefore, choice B would offer the most ideal conditions of those choices given for evaporation to take place. Choice D is a situation where evaporation is held to an absolute minimum if it exists at all.

100. C. For every one revolution made by the driver gear, the driven gear makes approximately two revolutions. This has the effect of increasing the speed of the drivetrain with corresponding sacrifice of torque. Choice A is true, but the question asked for the overall effect of the gear size disparity shown. This choice did not account for diminished torque. Choices B and D are false on their own merit.

TEST RATINGS ARE AS FOLLOWS

95–100 correct	EXCELLENT
87–94 correct	VERY GOOD
81–86 correct	GOOD
75–80 correct	FAIR
74 or below correct	UNSATISFACTORY

Go back to the questions you missed and determine if the question was just misinterpreted for one reason or another, or if it reflects a particular weakness in subject matter. If it is a matter of misinterpretation, try reading the question more slowly while paying close attention to key words such as *not, least, except, without,* etc. If, on the other hand, you determine a weakness in a particular area, don't despair. That is what this book is for; to identify any area of weakness before you take the actual exam. Reread the material on the area of concern in this book. If you still feel a need for supplemental material, your local library is an excellent source.

PRACTICE EXAM 4

Time allotted: 2 hours

DIRECTIONS: Each question has four choices for answers, lettered A, B, C, and D. Chose the best answer and then, on the answer sheet provided, find the corresponding question number and darken with a soft pencil the circle corresponding to the answer you have selected.

PRACTICE EXAM 4

INSTRUCTIONS FOR QUESTIONS 1–16

STUDY THE FOLLOWING FLOOR PLAN FOR ONLY 5 MINUTES. When your time is up, turn to the next page to answer questions 1 through 16 relating to the floor plan you have studied. YOU MAY NOT LOOK BACK AT THE FLOOR PLAN. NOTES ARE PROHIBITED.

REMEMBER: This is a memory exercise, so any review of this floor plan is prohibited on the actual examination. Thus, any review on the practice examination will skew your test results.

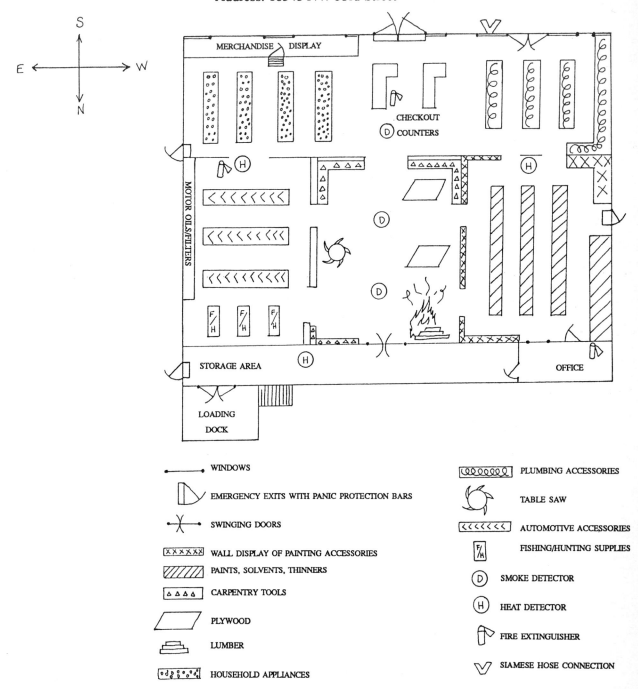

267

1. The floor plan diagrammed is that of a
    A. Clothing Store   B. Church   C. Hardware store   D. Tenement residence

2. Judging by the location of emergency exits in this floor plan, what can be assumed?
    A. This building shares a common partition with another structure on its west side.
    B. This building shares a common partition with two other buildings, one on the east side and the other on the west side.
    C. This structure shares the south wall with another business.
    D. This structure is free standing.

3. What is the address for the building diagrammed?
    A. 10543 NW 53rd Street
    B. 15043 NW 35th Street
    C. 10435 NE 51st Street
    D. 15043 NW 53rd Street

4. Where is the fire located in the diagram?
    A. The northeast side of the building
    B. The north side of the building
    C. The southeast side of the building
    D. The west side of the building

5. Which showroom does not offer an emergency exit that leads directly out of the building?
    A. Plumbing   B. Automotive   C. Household appliance   D. Painting accessories

6. Which alarm would probably be the first to respond to the fire?
    A. The heat detector in the lumber area
    B. The smoke detector in the plumbing area
    C. The heat detector in the storage area
    D. The northernmost smoke detector in the store

7. How many emergency exits (doors) are equipped with panic protection bars?
    A. 2   B. 3   C. 5   D. 7

8. Where is the loading dock with respect to the main entryway for customers?
    A. Northeast   B. North   C. Northwest   D. Southwest

9. If a customer were in the fishing and hunting department of the store when the fire broke out, what would be his or her safest means of escape?
    A. Out through the storage area to the loading dock
    B. Make a diagonal approach through the southeast section of the lumber department to the main entry on the south wall.
    C. Go directly to the paint department and use the exit provided there.
    D. Stay to the east side of the building as much as possible while getting to the exit provided in the household appliance department.

10. How many exterior windows does this building have?
    A. 5   B. 4   C. 3   D. 2

11. Where is the closest fire extinguisher in relation to the fire?
    A. The one in the automotive section
    B. The one in the paint department
    C. The one in the office
    D. The one in the storage area

12. The Siamese hose connection is nearest to what merchandise department?
    A. Household appliance
    B. Fishing and hunting supply
    C. Plumbing
    D. Clothing

13. What corner of the building is dedicated to merchandise display to entice people into the store?
    A. Southeast   B. Southwest   C. Northeast   D. Northwest

14. How many fire extinguishers are there in the store?
    A. 5   B. 4   C. 3   D. 2

15. Which area within the store is least prone to smoke damage due to the fire's location (assuming the fire was quickly suppressed)?
    A. The automotive department
    B. The storage area
    C. The paint department
    D. The checkout area

16. Where is the store's office in relation to the fishing and hunting department?
    A. West   B. East   C. North   D. South

17. Any time there is fire above the ground floor of a multistory building, empty hose line should be carried to the floor affected and then charged with water when the nozzle is in close vicinity to the fire. What is the obvious reason for this practice?
    A. Hose line can be checked for leaks prior to being charged.
    B. An empty hose line is significantly lighter and more maneuverable than a charged line.
    C. It saves wear and tear on the hose's exterior lining.
    D. There is a better chance of it not becoming entangled with an obstruction.

18. If a fire officer noticed that water flowing away from an immense fire was quite cold, what could be assumed?
    A. He must find an alternate source of water.
    B. There is a leak in the water main.
    C. The water streams being applied are not actually reaching the fire.
    D. The fire is almost extinguished.

19. A firefighter found it necessary to stretch a hose line through the back doors of a building. If the doors were made of tempered glass inserts enclosed in a steel framework, what would be the first thing done without reservation?
    A. Break a small opening near the lock and unlock the door from the inside.
    B. Insert a crowbar between the doors and twist gently.
    C. Use a battering ram to open the doors.
    D. Check to see if the doors are unlocked.

COMPLETE FIRE-FIGHTERS EXAM PREPARATION BOOK

20. Why is it generally more hazardous to break a closed and locked ordinary plate glass door panel than it is to break tempered glass panels of comparable size?
    A. Because ordinary plate glass breaks with dull edges
    B. Because ordinary plate glass breaks into extremely small pieces
    C. Because tempered glass will break into fairly large pieces
    D. Because ordinary plate glass will break into fairly large pieces with extremely sharp edges

21. Suppose you were on an aerial ladder or platform looking down on the house shown to the right. Which of the perspectives shown below would be the view probably seen?

    A. B. C. D.

22. Ground fires can be somewhat unpredictable because of directional changes of wind. However, if a fire is burning on a hillside where wind is not prevalent, fire will exhibit the tendency to burn more rapidly in which direction?
    A. Downhill
    B. Uphill
    C. In climates of higher humidity
    D. In rough terrain

ANSWER QUESTIONS 23 THROUGH 26 ON THE BASIS OF THE CHART BELOW

Chemical	Upper Flash Point	Classification	Lower Flash Point
Paint	100° F	Orange	
Sulfuric acid	182° F	Red	
Methyl vinyl ether	- 68.8° F	Green	
Pentane	- 40° F	Green	-129.8°F
Hexanone	57.2° F	Yellow	
Ethylbutylamine	64° F	Yellow	
Pentanediol	275° F	Red	- 15.6° F
Dihexylamine	- 220° F	Blue	
Monochloromethane	32° F	Yellow	- 97.7° F
Acetate	190° F	Red	

270

23. Which chemical presents the greatest danger to those who handle it with respect to its susceptibility to ignite?
    A. Pentane
    B. Methyl vinyl ether
    C. Pentanediol
    D. Monochloromethane

24. Which chemical demonstrates the widest flash point variance?
    A. Acetate   B. Monochloromethane   C. Pentanediol   D. Pentane

25. Which chemical belongs to a red classification and has a flash point greater than 182° F?
    A. Sulfuric acid   B. Hexanone   C. Dihexylamine   D. Acetate

26. Which of the following assumptions could be made about the chemicals listed in this chart?
    A. Those chemicals that are classified as green have the lowest flash points.
    B. Those chemicals that are classified as yellow have a fairly narrow upper flash point variance ranging from freezing to nearly 65° F.
    C. Those chemicals that demonstrate a red classification are fairly unstable.
    D. More chemicals are classified as being either green or yellow than any other two-color combination.

27. Of the following first aid principles, which is probably the most important to follow first before other procedures are conducted in attempting to save someone's life?
    A. Restore breathing
    B. Stop major bleeding
    C. Treat for shock
    D. Clean any superficial wounds

28. If a firefighter were ordered by a commanding officer to do something that would definitely endanger the lives of his co-workers, what is the best possible way for him to handle such situation?
    A. Assure his superior that he would do as asked since he was in charge and therefore not prone to making mistakes.
    B. Protest immediately to the battalion chief.
    C. Ask for an immediate transfer.
    D. Ask for a reclarification of the orders and then if he still feels the same, explain to the commanding officer the reason for refusing to comply with the orders.

29. Which of the following infractions committed by a firefighter is probably the most serious and warrants dismissal as a consequence?
    A. Being under the influence of any unauthorized controlled substance or drug possession while on duty.
    B. Disorderly conduct within the fire house.
    C. Inattentiveness, loafing or some other unauthorized activity while on duty.
    D. Failure to follow prescribed procedures and practices in the line of duty.

30. The angle of a hose stream attacking a structural fire has direct bearing on how far that stream will penetrate. According to the diagram below, which angle of attack demonstrates the greatest penetration of a hose stream? (Assume the nozzle is in a fixed position 30 feet from the base of the building).

        A.   B.   C.   D.

31. What is the reason that upper floor flooding as a firefighting tactic is used only as a last resort to control a fire directly below?
    A. It involves an inordinate amount of manpower.
    B. Water damage to the building involved can be fairly extensive.
    C. Water supply to other lines may be interrupted.
    D. It is a very ineffective technique in combating multilevel structural fires.

32. Most building codes require that any structural steel in high-rise buildings be covered by fire-resistant material such as gypsum block, vermiculite plaster, or concrete. What can be concluded from this statement?
    A. Fire codes can be extremely strict.
    B. Structural steel is covered for purely aesthetic reasons.
    C. The coverings mentioned can serve to prop up structural weakness caused by fire.
    D. Structural steel is not very fire resistant.

33. When firefighters arrived at the scene of a third-floor fire in a four-story building, it was imperative that hose lines be brought up through the side stairwells to the affected floor. Three lengths of coupled hose allowed for plenty of reserve line. But before the line became charged, one of the firefighters took this reserve and ran a loop up to the fourth floor. What would this practice be considered?
    A. Efficient because it is much easier to pull a charged hose line down a stairwell than to pull it up one.
    B. Efficient because the reserve hose line laid in this fashion presents a lesser chance of being stepped on by co-workers.
    C. Inefficient because it requires the pump to work harder to send water that much higher.
    D. Inefficient because it increases the chance of a hose line to kink.

34. Ship fires present unique problems to firefighters trying to contain a blaze that originates in a cargo hold, engine room, storage lockers, or living area. Which problem below is the least likely to be encountered by firefighters?
    A. Limited access to the area affected.
    B. Fire may have to be approached from the top down.
    C. Water supply limitations.
    D. If the fire is fairly intense, flooding may be required, which can affect the seaworthiness of the vessel involved.

35. It has been found that in the majority of home fires, smoke detectors proved to be worthless as an early warning system for the residents due to what fact?
    A. Smoke detectors were incorrectly positioned within the home.
    B. Many earlier model designs were defective.
    C. Homeowners neglected to periodically check the batteries.
    D. Smoke detectors were deliberately disabled by the homeowner because of false alarms.

36. Providing that wind and open flame exposure are factors to consider, which of the following approaches to a burning structure would be best with respect to the use of an aerial ladder?
    A. Position the ladder truck north of the structure.
    B. Position the ladder truck south of the structure.
    C. Position the ladder truck downhill from the structure.
    D. Position the ladder truck uphill from the structure.

37. In the aftermath of extinguishing a convenience store fire, firefighter Jim Tyrell was directed to assist with overhaul and salvage operations. While doing so, Mr. Tyrell noticed a co-worker pick something up from the burned debris and place the item in his coat pocket. What is the most appropriate action for Mr. Tyrell to take in this situation?
    A. Ignore it because burned items are practically worthless and are covered by insurance anyway.
    B. Ignore it because turning in a co-worker can be detrimental to the inherent trust that exists between firefighters.
    C. Ascertain the facts by confronting the co-worker before any conclusions are drawn. Only then should the event be detailed to a supervisor for handling if any rules or regulations have been violated.
    D. Immediately notify his superior that a theft has taken place.

38. Firefighting tactics can be directly influenced by one or more factors. The type of fire, size of the fire, location of the fire, and the extension probability for that fire need to be accounted for in formulating effective firefighting strategies. From a safety standpoint, which of these four factors is probably the most important consideration for firefighting personnel?
    A. Type of fire
    B. Location of the fire
    C. Size of the fire
    D. Probability of extension

39. If firefighters in a rural area were forced to rely on gravitational water flow from a ground-level water storage tank, which of the following statements would be true with regard to flow velocity via the tank's discharge port?
    A. The smaller the volume of water within the tank, the higher the flow velocity.
    B. There is no discernable difference of flow velocity regardless of how full the water tank is.
    C. The larger the volume of water within the tank, the lower the flow velocity.
    D. The larger the volume of water within the tank, the higher the flow velocity.

40. The time at which a fire occurs can have direct bearing on the potential threat it poses to individuals. Which of the time periods below is probably the worst in this regard?
    A. 8 a.m. to 4 p.m.
    B. 4 p.m. to 12 midnight
    C. 12 midnight to 8 a.m.
    D. A fire at any time of day or night poses the same kind of threat to life and property.

41. Despite the fact that schools regularly conduct fire drills, school fires warrant special concerns with regard to safe evacuation. What is the most probable reason for this?
    A. There is usually a lot of confusion with regard to whether or not everyone made it out of the building safely.
    B. Children have the inherent tendency to hide from a fire rather than seek a safe escape to the outside of the building.
    C. Children often behave irrationally during a fire.
    D. All of the above.

42. The formula for translating Fahrenheit temperature readings into Celsius temperature readings is $(F° - 32) \, 5/9$. What would 99° Fahrenheit be if converted to Celsius?
    A. 81.3°   B. 67.8°   C. 41.2°   D. 37.2°

43. How many milliliters are there in 12.38 liters?
   A. 1238 B. 123,800 C. 12,380 D. .001238

44. Quite often metrics need to be converted to standard measures. One such example could involve changing kilometers into miles. Assuming there are 1.6 kilometers per mile, what would be the distance in miles between two towns that are 96 kilometers apart?
   A. 5 miles B. 102.3 miles C. 60 miles D. 175 miles

45. If a mechanic measures the diameter (i.e., the pitch) of a gear with a micrometer and figures that it is 2.5% larger than the manufacturer's specifications of 5.735 inches, what is the actual pitch of the gear in question?
   A. 5.88 inches B. 6.33 inches C. 6.721 inches D. 7.21 inches

46. If a rectangular area of 47.5 feet by 13.2 feet were roped off for arson investigation, how many square feet would this encompass?
   A. 627 square feet
   B. 593 square feet
   C. 572.5 square feet
   D. 402 square feet

47. A chemical storage tank is cylindrical, with a height of 32.5 feet and a diameter of 18.5 feet. If the acetyl chloride in this tank were at the 20.7 foot mark, how many gallons of acetyl chloride are there, assuming there are 7.48 gallons per cubic foot?
   A. 40,640.51 gallons
   B. 41,620.404 gallons
   C. 43,790.2 gallons
   D. 50,000 gallons

48. The rectangle and square shown below have the same area. On the basis of the information given, determine the length of the square's side.

   A. 16.37 meters B. 17.19 meters C. 17.25 meters D. 18.88 meters

49. The right triangle shown has an area of 157.6 square meters. If there is a 4:3 relationship existing between the base and the height of this triangle, what is the length of side BC?

   A. 23.12 meters B. 22.713 meters

   C. 20.5 meters D. 15.375 meters

PRACTICE EXAM 4

50. A fire truck had a 14-quart coolant system that contained a 45% antifreeze solution. Approximately how many quarts of this solution would have to be drained and replaced with pure antifreeze to achieve a 75% antifreeze solution?

    A. 5.23 quarts B. 7.64 quarts C. 8.91 quarts D. 9 quarts

51. Class A fire extinguishers can protect a maximum area of 557 square meters. A commercial structure had 132,000 square feet of floor space, and the owner wanted to adequately protect this building with Class A fire extinguishers. What would be the required number of extinguishers needed? (Assume there are 10.764 ft² per meter²).

    A. 22.02 B. 23 C. 23.4 D. 24

52. The cylindrical can shown has a diameter of 22 inches and a height of 66 inches. Assuming there are 231 cubic inches per gallon, how many gallons could this container hold?

    A. 108.61 gallons   B. 106.8 gallons

    C. 103.75 gallons   D. 98.93 gallons

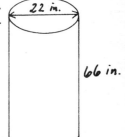

53. Pulley A has a 13-inch diameter
    Pulley B has a 9-inch diameter
    Pulley C has a 16-inch diameter
    Pulley D has a 6-inch diameter
    (Pulleys B and C turn on the same shaft).
    If Pulley A rotates at 54 rpm, how fast does
    Pulley D rotate?

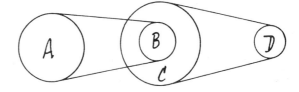

    A. 407 rpm B. 362 rpm C. 208 rpm D. 139 rpm

54. If an aluminum alloy bar measured 4 feet by 2 inches by 2 inches and weighed 12.3 pounds, how much would a bar of the same material measuring 7 feet by 3 inches by 4 inches weigh?

    A. 64.57 pounds B. 56.05 pounds C. 43.757 pounds D. 26.12 pounds

ANSWER QUESTIONS 55 THROUGH 60 ON THE BASIS OF THE PASSAGE BELOW

Many deaths each year can be attributed to carbon monoxide poisoning. Since carbon monoxide is colorless and almost odorless, few people realize its presence before it is too late. Carbon monoxide, or CO as it is known chemically, is the product of incomplete combustion of various fuels. Carbon monoxide is lethal when it is inhaled by the victim because it readily displaces oxygen from the blood. In fact, as little as .0007 or 700 parts per million of carbon monoxide in the air is enough to displace half of the oxygen present in the blood. Anything greater than 1% is enough to kill most people. Various factors can determine an individual's susceptibility to carbon monoxide poisoning. Someone who is a child or petite in stature, or exercising, is much more prone to being overcome by the gas. Physical ailments such as emphysema, heart disease, asthma, bronchitis, and others also serve to enhance susceptibility.

Symptomatically, carbon monoxide asphyxiation may manifest itself in a variety of ways. Extreme headaches, dizziness, nausea, weakness, and increased pulse and respiration are preliminary to collapse and loss of consciousness. As an individual draws nearer to death, reflexes are inhibited, muscular control is forfeited, blood pressure drops, and finally breathing stops.

Carbon monoxide is not physiologically used or consumed by the body. Therefore, the only way it can be eliminated from the body is to administer pure oxygen. This process may be expedited with the use of a hyperbaric chamber.

55. What is the main principle of this reading?
    A. Only a small amount of carbon monoxide need be inhaled to cause asphyxiation in most people.
    B. Carbon monoxide is a product of complete combustion.
    C. A short overview of carbon monoxide, its origin and lethal characteristics.
    D. What to expect in the way of symptoms after inhaling carbon monoxide.

56. Carbon monoxide is chemically known as:
    A. CO_2 B. CO C. O_2 D. COC_2

57. What concentration of carbon monoxide needs to be present to cause death for most victims?
    A. 0.007   B. 0.0007   C. .1   D. 0.01

58. On the basis of this article, which of the following people described below would be more prone to carbon monoxide asphyxiation?
    A. A 160 pound, 5'11" tall male who smokes.
    B. A 190 pound, 6'3" tall male allergic to formaldehyde.
    C. A 95 pound, 5'5" tall female.
    D. A 125 pound, 5'9" tall female who suffers from a cold.

59. Which of the following circumstances is more conducive to carbon monoxide poisoning?
    A. A barbecue grill used to heat a room during a cold spell.
    B. Standing in close proximity to a muffler of a car idling in a carport.
    C. People huddling around a campground fire to stay warm and roast marshmallows.
    D. The use propane gas in various household appliances to heat or cook with.

60. All of the symptoms given below can be attributed to carbon monoxide poisoning, with the exception of which one?
    A. Severe headache
    B. The desire to vomit
    C. An initial decrease of breathing rate
    D. Loss of strength

ANSWER QUESTIONS 61 THROUGH 65 ON THE BASIS OF THE PASSAGE BELOW

A recent innovation that has improved safety for firefighters working in hazardous environments is a device called PASS or Personal Alert Safety System. This device is designed to emit an alarm when the firefighter wearing it does not move for 20 to 30 seconds. This essentially notifies other personnel that a firefighter has become unconscious or incapacitated for some reason. The alarm serves as a beacon for rescuers.

The device itself is simple to operate and is compact enough to be fastened to the outside of a jacket or harness strap. The switch on the unit has only three settings: ON, ARM, and OFF. When the device is switched to the ON position, a regular distress alarm is emitted. If the unit is on ARM, it will emit a short distress alarm to indicate that it is functioning properly. If a firefighter does not move for 20 to 30 seconds thereafter, a distress alarm will be emitted. Another advantage to this unit is that it is battery operated, thus reducing the possibility of a spark. This is particularly important if a firefighter is working in an environment containing flammable gases or explosive compounds. PASS is an indispensable tool that should be worn by all firefighters.

61. What is the acronym for the device described in the article?
    A. Personal Alert Safety System
    B. PASS
    C. Battery-operated unit
    D. Locational beacon

62. What did the ending of the article imply?
    A. Personal Alert Safety Systems are indispensable tools for firefighters.
    B. PASS is very affordable for most fire departments.
    C. The unit itself is convenient to use.
    D. Not all firefighting personnel wear PASS.

63. If a firefighter wanted to determine if his or her PASS had adequate battery charge, he or she would manually switch the unit to what position?
    A. ON   B. ARMed   C. ON or ARMed   D. OFF

64. In which of the following circumstances would PASS benefit a firefighter?
    A. If the firefighter was rendered unconscious.
    B. If the firefighter became pinned beneath fallen debris.
    C. If the firefighter became disoriented and extremely fatigued.
    D. All of the above.

65. Once the unit is armed, how long is it necessary for a firefighter to be immobilized before an alarm is sounded?
    A. 20–30 seconds   B. 20–30 minutes   C. 15–20 minutes   D. 15–20 seconds

ANSWER QUESTIONS 66 THROUGH 70 ON THE BASIS OF THE PASSAGE BELOW

It is an unfortunate fact that many homeowners tragically lose their lives because of apathy expressed toward home fire drills. EDITH, or Exit Drills In The Home, and Learn Not To Burn are two programs recently implemented by fire departments nationwide to counter such attitudes. Both of these programs, developed by the National Fire Protection Association (NFPA), accentuate the installation and maintenance of smoke detection equipment and teach users to establish an evacuation procedure for all occupants and conduct fire drills at regular intervals to practice what has been adopted. The NFPA recommends that any evacuation plan should include a minimum of two means of escape from every room in the household. If the residence is multistoried, fire escape ladders of appropriate length should be provided. The plan should also include some predetermined point outside the household where residents should meet and be accounted for. This meeting place should be a safe distance from the home and not at a point where they may encumber firefighters responding to the emergency. Fire inspection officers are more than willing to review any such plans with a homeowner and offer suggestions that will enhance fire safety.

It is the goal of NFPA that home fire safety programs make more people aware of what needs to be done prior to and during a fire. No doubt, over time, fire fatality statistics for residences will substantially decrease due to such programs.

66. What would be an appropriate title for this article?
    A. The Aspirations of NFPA
    B. The Implementation of Home Fire Safety Plans and Their Related Benefits
    C. The Consequences of Home Fire Safety Apathy
    D. The History of EDITH and Learn Not to Burn Programs

67. What is EDITH an acronym for, according to the article?
    A. The name of the person who developed the program.
    B. Emergency Dispatch in Time to Help
    C. Exit Drills in the Home
    D. Emergency Doors In the Home

68. According to the reading, who was responsible for developing the two home fire safety programs that are now nationally recognized?
    A. Edith
    B. National Association of Firefighters
    C. International Fire Service Training Association
    D. National Fire Protection Association

69. Which of the following was not discussed as being a recommended procedure according to EDITH and Learn Not To Burn home fire safety programs?
    A. Assemble at one point within the home so that everyone can be accounted for.
    B. Be sure that smoke detectors in the home are functioning properly.
    C. Have an evacuation plan prepared and understood by all residents.
    D. Conduct fire drills on a regular basis.

70. If a master bedroom in a split-level home had two means of egress, one door leading to a hallway and another door leading to a master bathroom that is connected to an adjoining bedroom, what, if anything, should be done to further enhance home fire safety according to the reading?
    A. Since the master bedroom already has two means of egress, no further action is required.
    B. Install panic protection bars on both doors.
    C. Install emergency lighting by the front door.
    D. Install a short ladder immediately below one of the master bedroom windows.

ANSWER QUESTIONS 71 THROUGH 77 ON THE BASIS OF THE PASSAGE BELOW

Electrical flow can be attributed to the movement or transfer of electrons between atoms. If this process takes place readily within a given material, that material is considered to be conductive or serves as a good electrical conductor. Examples of excellent conductors are gold, silver, or copper. Because of its relative low cost compared to other such materials, copper is used more widely in electrical projects.

If the movement of electrons is greatly inhibited by a material, the material is considered to be nonconductive or an insulator. Rubber, glass, and most plastics are three examples of nonconductors or insulators.

When the movement of electrons is only partially restricted in a material, with its flow being intermediate between that of a nonconductor and a conductor, the material is considered semiconductive. Silicon and germanium are examples of semiconductors.

Electrical energy flow through a conductor is comparable to how water flows through a system. The three factors that affect water flow are pressure, measured in pounds per square inch (psi); resistance, or frictional loss; and quantity, which is measured in liters or gallons per minute.

Electricity is also measured by three factors; however, the terms are different. When considering electrical flow, electrical pressure is measured in volts. Resistance is measured in ohms, and quantity of electricity (current) is expressed in amperes. A correlation worth remembering is that if electrical flow through a material is marginal, the resistance in ohms is greater. This would be the case if a material used for an electrical pathway was a poor conductor.

Each of these three factors directly relates to the others in a quantitative sense, too. According to Ohm's law, the electrical pressure in volts is equal to the resistance in ohms multiplied by the current in amperes. If one of these factors is unknown, it can be solved for algebraically because of the relationship that exists among the three factors. Once these basics are understood, the concept or principles of electricity are easier to comprehend.

71. What is the basic intent of this article?
    A. To illustrate the differences between water flow and electrical energy flow.
    B. To highlight what makes a material a good conductor versus a poor conductor.
    C. To lend insight into how various factors that affect electrical energy flow are quantitatively interrelated and can be determined by Ohm's law.
    D. To simplify basic electrical principles by comparing them to water flow.

72. What part of an atom is responsible for electrical energy transfer as discussed in the reading?
    A. Electrons   B. Protons   C. Neutrons   D. It was not mentioned in the reading.

73. On the basis of this article, we could probably assume that wood is a
    A. Good conductor
    B. Excellent conductor
    C. Poor conductor
    D. Semi-conductor

74. If an electrician noticed that a particular electrical circuit had a high ohm reading, he or she could assume which of the following?
    A. The electrical pathway in the circuit presented a low degree of resistance.
    B. The electrical pathway in the circuit presented a high degree of resistance.
    C. Friction loss is minimal
    D. The wire used was an excellent conductor.

75. Liters per minute is to water flow as _____ is to electrical energy flow.
    A. Amperes   B. Voltage   C. Ohms   D. None of the above

76. Which of the following relationships is correct according to Ohm's law as described in the reading?
    A. If the voltage variable doubles, ohm and amperes variables will double as well.
    B. If the voltage variable triples, ohm and amperes variables will quadruple.
    C. If voltage remains constant, and resistance increases, amperes or current proportionally decreases.
    D. If amperes and voltage variables remain constant, the resistance variable must double.

77. Why is copper used as a conductor in more industrial applications than gold or silver?
    A. It is a better conducting material.
    B. It is more readily available.
    C. It is a heavier material that will endure the elements better.
    D. It remains relatively cooler when subjected to large electrical-energy flow.

ANSWER QUESTIONS 78 THROUGH 87 ON THE BASIS OF THE MAP BELOW

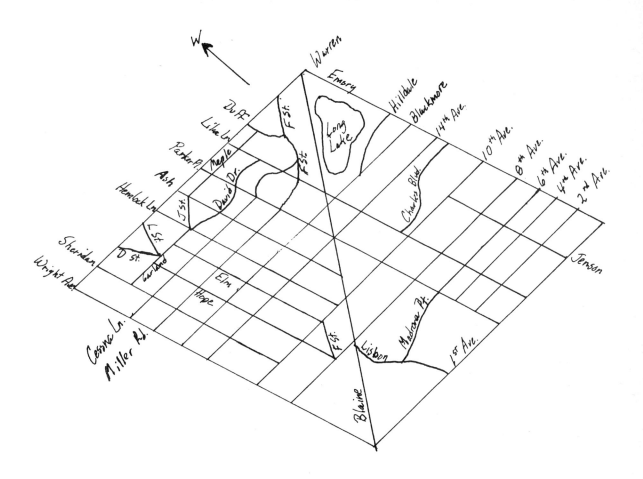

Four fire district jurisdictions cover the area shown in the map above.

Fire Station #17 is responsible for the area bounded by Emory from Warren to 14th Avenue, 14th Avenue to Parker Place, Parker Place to Miller Road, Miller Road to F Street, and then F Street to Warren.

Fire Station #9 is responsible for the area bounded by F Street from Warren to Miller Road, Miller Road to Parker Place, Parker Place to 14th Avenue, 14th Avenue to Hemlock Lane, Hemlock Lane to Garland, Garland to Sheridan Place, Sheridan Place to Warren, and then Warren back to F Street.

Fire Station #13 is responsible for that area bounded by Sheridan Place from Warren to Garland, Garland to Hemlock Lane, Hemlock Lane to 14th Avenue, 14th Avenue to Blaine, Blaine to Wright Avenue, Wright Avenue to Warren and then back to Sheridan Place on Warren.

Fire Station #5 is responsible for that area bounded by 14th Avenue from Emory to Blaine, Blaine to 1st Avenue, 1st Avenue to Emory, and then Emory back to 14th Avenue.

281

78. If a two-car accident occurred at the intersection of Duff and Maple, what fire station does it concern?
    A. #5   B. #9   C. #13   D. #17

79. The block bounded by 14th Avenue, Ash, and Blaine streets falls under which fire station's jurisdiction?
    A. #5   B. #9   C. #13   D. #17

80. Considering the boundaries of Fire Station #5's jurisdiction, if the fire station was said to be centrally located with respect to those boundaries, we can assume that Fire Station #5 is where?
    A. On the northwest corner of the block bounded by 6th and 8th Avenues, Lilac Lane, and Parker Place.
    B. On the southwest corner of the block bounded by 8th and 10th Avenues, Jenson, and Lilac Lane.
    C. On the southeast corner of the 14th Avenue and Charles Boulevard intersection
    D. On the northeast corner of the block bounded by 4th and 6th Avenues, Jenson, and Lilac Lane.

81. Of the four fire stations given, which one serves the smallest area and perhaps the fewest people?
    A. #5   B. #9   C. #13   D. #17

82. If a vehicle were heading south on 14th Avenue, and made a left turn three blocks past Ash and then went two blocks east and then one more block south before heading west four more blocks, where would this vehicle end up?
    A. The intersection of Sheridan Place and Garland
    B. The intersection of Miller Road and Wright Avenue
    C. The intersection of Hope and 8th Avenue
    D. The intersection of Blackmore and Sheridan Place

83. Where is the intersection of 8th Avenue and Jenson in relation to the intersection of 14th Avenue and Parker Place?
    A. Northwest   B. North   C. Northeast   D. Southwest

84. What direction does F Street run from 8th Avenue?
    A. Northwest   B. Southeast   C. West   D. East

85. A residence located on the northeast corner of the 14th Avenue and Parker Place intersection would be within which fire station's jurisdiction?
    A. #5   B. #9   C. #13   D. #17

86. If Fire Station #13 were located on the southwest corner of the Hope and Blackmore street intersection, what would be the most direct approach to a residential fire located on the northwest corner of 10th Avenue and Hemlock Lane?
    A. Go west two blocks from the station and then north two blocks.
    B. Go south two blocks from the station and then west two blocks.
    C. Go east two blocks from the station and then south one block.
    D. Go east two blocks from the station and then north two blocks.

87. If a fire apparatus were heading north on 10th Avenue and made a left onto Lilac Lane and went two blocks before making another left turn, in what direction would this vehicle be heading at that point?
    A. North   B. South   C. East   D. West

88. Which of the following statements is true with respect to the drivetrain illustrated below?

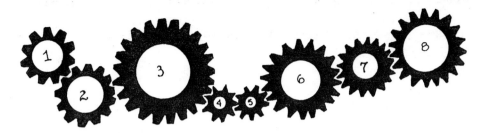

   A. Gears 2 and 8 will turn clockwise if the first gear turns clockwise.
   B. Gears 4 and 5 will turn counterclockwise if Gear 2 turns clockwise.
   C. Gears 3, 5, and 6 will turn in the same direction that Gear 2 turns.
   D. Gears 3, 5, and 7 will turn counterclockwise if Gear 2 turns clockwise.

ANSWER QUESTIONS 89 THROUGH 91 ON THE BASIS OF THE DIAGRAM BELOW

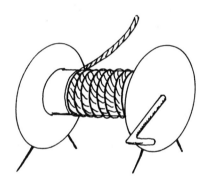

89. What is the one feature most responsible for the leverage gained by using the windlass illustrated?
   A. The diameter of rope utilized on the drum.
   B. The length of the handgrip (i.e., that portion of the handle that extends in a horizontal axis away from the drum).
   C. The length of handle (i.e., that portion of the handle extending perpendicular to the drum) in relation to the drum's diameter.
   D. The length of the drum itself.

90. If 400 pounds of tension had to be applied to the line and the windlass had a drum diameter of 6 inches, how long a handle would be necessary if only 125 pounds of potential effort is available?
   A. 60.32 inches   B. 51.7 inches   C. 25.1 inches   D. 19.2 inches

91. From the information given in question 90, what is the mechanical advantage gained by using this particular windlass?
   A. 4.7 : 1   B. 3.2 : 1   C. 2.8 : 1   D. 1.9 : 1

92. Two firefighters were assisting in a toxic waste cleanup. If both firefighters attempted to roll a 55-gallon drum of chemicals weighing 460 pounds up the inclined plane shown to the right in order to load it onto a disposal truck, how much effort is needed from each firefighter?

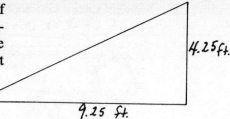

A. 192 pounds B. 96 pounds

C. 138 pounds D. 69 pounds

93. With all things considered, what kind of lift advantage is gained under the circumstances described in question 92 versus an individual trying to lift the barrel outright?
A. 2.4 : 1   B. 3.33 : 1   C. 4.8 : 1   D. 5.7 : 1

94. How much effort would be required using this pair of pulleys set up in the manner diagrammed to lift the weight shown? (Assume pulley B is in a fixed position).

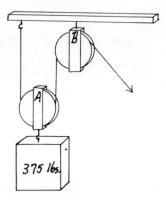

A. 187.5 pounds B. 205.5 pounds

C. 250.5 pounds D. 262.7 pounds

95. Referring to the diagram in question 94, if the cable in this pulley apparatus were pulled 15 feet (assuming there was enough cable present to do so), how high would this lift the weight?
A. 7.5 feet   B. 10 feet   C. 15 feet   D. 17.5 feet

96. Which of the following statements is true with regard to the belt drive illustrated below? (NOTE: Wheels 3 and 4 are keyed to the same shaft.)
Wheel dimensions are as follows:
1. : 8 inches in diameter
2. : 4 inches in diameter
3. : 4 inches in diameter
4. : 8 inches in diameter
5. : 4 inches in diameter

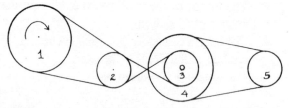

A. Wheel 4 will turn faster than wheel 2 in a counterclockwise direction.
B. Wheel 5 will turn faster than wheel 1 in a clockwise direction.
C. Wheel 2 and wheel 5 will turn in the same direction at the same speed.
D. Wheel 3 will turn faster than wheel 1 in a counterclockwise direction.

97. If the jackscrew illustrated to the right is needed to lift 2½ tons and it is known the jack has a ⅛-inch pitch and a handle/lever measuring 19 inches in length, how much effort is required to achieve lift?

   A. 3.15 pounds   B. 5.24 pounds

   C. 7.9 pounds   D. 13.71 pounds

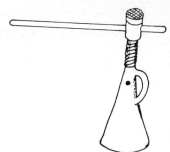

98. If a drivegear has 28 teeth and a connecting gear (i.e., the driven gear) has 14 teeth, which of the following events will occur?
   A. The torque output and speed are doubled.
   B. The torque output is only ¼ of that input while the speed is tripled.
   C. The torque output is doubled while the speed is halved.
   D. The torque output is halved while the speed is doubled.

99. Study the illustration to the right. Considering the fact that gears A and C are 50% the size of gears B and D in the driveline, what is the overall advantage gained by using such a gear configuration?

   A. Speed enhancement on the order of 4:1.
   B. Speed enhancement on the order of 2:1.
   C. Torque enhancement on the order of 4:1.
   D. Torque enhancement on the order of 2:1.

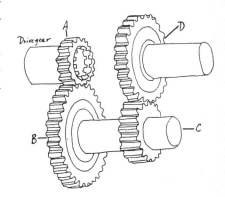

100. Cylinders 1, 2, and 3 have 5-, 7-, and 13-liter capacities, respectively. How much water would have to be pumped through the inlet port to raise the level of cylinder 3 from 4.5 liters to 9.75 liters?

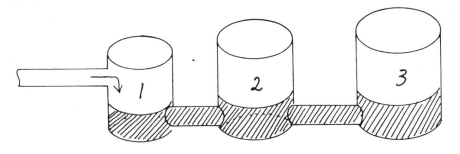

   A. 10.1 liters   B. 8.7 liters   C. 5.25 liters   D. 2.423 liters

ANSWER SHEET TO EXAM 4

1. Ⓐ Ⓑ Ⓒ Ⓓ
2. Ⓐ Ⓑ Ⓒ Ⓓ
3. Ⓐ Ⓑ Ⓒ Ⓓ
4. Ⓐ Ⓑ Ⓒ Ⓓ
5. Ⓐ Ⓑ Ⓒ Ⓓ
6. Ⓐ Ⓑ Ⓒ Ⓓ
7. Ⓐ Ⓑ Ⓒ Ⓓ
8. Ⓐ Ⓑ Ⓒ Ⓓ
9. Ⓐ Ⓑ Ⓒ Ⓓ
10. Ⓐ Ⓑ Ⓒ Ⓓ
11. Ⓐ Ⓑ Ⓒ Ⓓ
12. Ⓐ Ⓑ Ⓒ Ⓓ
13. Ⓐ Ⓑ Ⓒ Ⓓ
14. Ⓐ Ⓑ Ⓒ Ⓓ
15. Ⓐ Ⓑ Ⓒ Ⓓ
16. Ⓐ Ⓑ Ⓒ Ⓓ
17. Ⓐ Ⓑ Ⓒ Ⓓ
18. Ⓐ Ⓑ Ⓒ Ⓓ
19. Ⓐ Ⓑ Ⓒ Ⓓ
20. Ⓐ Ⓑ Ⓒ Ⓓ
21. Ⓐ Ⓑ Ⓒ Ⓓ
22. Ⓐ Ⓑ Ⓒ Ⓓ
23. Ⓐ Ⓑ Ⓒ Ⓓ
24. Ⓐ Ⓑ Ⓒ Ⓓ
25. Ⓐ Ⓑ Ⓒ Ⓓ
26. Ⓐ Ⓑ Ⓒ Ⓓ
27. Ⓐ Ⓑ Ⓒ Ⓓ
28. Ⓐ Ⓑ Ⓒ Ⓓ
29. Ⓐ Ⓑ Ⓒ Ⓓ
30. Ⓐ Ⓑ Ⓒ Ⓓ
31. Ⓐ Ⓑ Ⓒ Ⓓ
32. Ⓐ Ⓑ Ⓒ Ⓓ
33. Ⓐ Ⓑ Ⓒ Ⓓ
34. Ⓐ Ⓑ Ⓒ Ⓓ
35. Ⓐ Ⓑ Ⓒ Ⓓ
36. Ⓐ Ⓑ Ⓒ Ⓓ
37. Ⓐ Ⓑ Ⓒ Ⓓ
38. Ⓐ Ⓑ Ⓒ Ⓓ
39. Ⓐ Ⓑ Ⓒ Ⓓ
40. Ⓐ Ⓑ Ⓒ Ⓓ
41. Ⓐ Ⓑ Ⓒ Ⓓ
42. Ⓐ Ⓑ Ⓒ Ⓓ
43. Ⓐ Ⓑ Ⓒ Ⓓ
44. Ⓐ Ⓑ Ⓒ Ⓓ
45. Ⓐ Ⓑ Ⓒ Ⓓ
46. Ⓐ Ⓑ Ⓒ Ⓓ
47. Ⓐ Ⓑ Ⓒ Ⓓ
48. Ⓐ Ⓑ Ⓒ Ⓓ
49. Ⓐ Ⓑ Ⓒ Ⓓ
50. Ⓐ Ⓑ Ⓒ Ⓓ
51. Ⓐ Ⓑ Ⓒ Ⓓ
52. Ⓐ Ⓑ Ⓒ Ⓓ
53. Ⓐ Ⓑ Ⓒ Ⓓ
54. Ⓐ Ⓑ Ⓒ Ⓓ
55. Ⓐ Ⓑ Ⓒ Ⓓ
56. Ⓐ Ⓑ Ⓒ Ⓓ
57. Ⓐ Ⓑ Ⓒ Ⓓ
58. Ⓐ Ⓑ Ⓒ Ⓓ
59. Ⓐ Ⓑ Ⓒ Ⓓ
60. Ⓐ Ⓑ Ⓒ Ⓓ
61. Ⓐ Ⓑ Ⓒ Ⓓ
62. Ⓐ Ⓑ Ⓒ Ⓓ
63. Ⓐ Ⓑ Ⓒ Ⓓ
64. Ⓐ Ⓑ Ⓒ Ⓓ
65. Ⓐ Ⓑ Ⓒ Ⓓ
66. Ⓐ Ⓑ Ⓒ Ⓓ
67. Ⓐ Ⓑ Ⓒ Ⓓ
68. Ⓐ Ⓑ Ⓒ Ⓓ
69. Ⓐ Ⓑ Ⓒ Ⓓ
70. Ⓐ Ⓑ Ⓒ Ⓓ
71. Ⓐ Ⓑ Ⓒ Ⓓ
72. Ⓐ Ⓑ Ⓒ Ⓓ
73. Ⓐ Ⓑ Ⓒ Ⓓ
74. Ⓐ Ⓑ Ⓒ Ⓓ
75. Ⓐ Ⓑ Ⓒ Ⓓ
76. Ⓐ Ⓑ Ⓒ Ⓓ
77. Ⓐ Ⓑ Ⓒ Ⓓ
78. Ⓐ Ⓑ Ⓒ Ⓓ
79. Ⓐ Ⓑ Ⓒ Ⓓ
80. Ⓐ Ⓑ Ⓒ Ⓓ
81. Ⓐ Ⓑ Ⓒ Ⓓ
82. Ⓐ Ⓑ Ⓒ Ⓓ
83. Ⓐ Ⓑ Ⓒ Ⓓ
84. Ⓐ Ⓑ Ⓒ Ⓓ
85. Ⓐ Ⓑ Ⓒ Ⓓ
86. Ⓐ Ⓑ Ⓒ Ⓓ
87. Ⓐ Ⓑ Ⓒ Ⓓ
88. Ⓐ Ⓑ Ⓒ Ⓓ
89. Ⓐ Ⓑ Ⓒ Ⓓ
90. Ⓐ Ⓑ Ⓒ Ⓓ
91. Ⓐ Ⓑ Ⓒ Ⓓ
92. Ⓐ Ⓑ Ⓒ Ⓓ
93. Ⓐ Ⓑ Ⓒ Ⓓ
94. Ⓐ Ⓑ Ⓒ Ⓓ
95. Ⓐ Ⓑ Ⓒ Ⓓ
96. Ⓐ Ⓑ Ⓒ Ⓓ
97. Ⓐ Ⓑ Ⓒ Ⓓ
98. Ⓐ Ⓑ Ⓒ Ⓓ
99. Ⓐ Ⓑ Ⓒ Ⓓ
100. Ⓐ Ⓑ Ⓒ Ⓓ

ANSWERS TO PRACTICE EXAM 4

Refer to the floor plan illustration for any clarification on Questions 1 through 16.

1. C	7. C	13. A
2. D	8. A	14. C
3. A	9. D	15. B
4. B	10. B	16. A
5. B	11. C	
6. D	12. C	

17. *B.* A charged hose line is significantly heavier than an empty line. Not only would it be more cumbersome for firefighters to lug up a flight of stairs, but it is relatively stiff and difficult to maneuver. Of the other choices given, choice C is probably true since the line is not drawn over various obstacles. However, it is not the main reason for the practice described in the question. Choice A is false because hose line should be checked for leaks prior to a fire. In fact, the last time the hose was used, any leaks should have been marked and repaired immediately thereafter. Choice D is false because uncharged line can inadvertently become kinked or pinched by an obstruction whereas a charged line will not, because of its stiffness.

18. *C.* If a hose stream is incorrectly applied to a fire, heat absorption is minimized, thus nullifying the affect that water should have on a fire. The other choices given are false on their own merit.

19. *D.* Before any forcible entry is attempted, the door(s) should first be checked to see if they are in fact locked. Using a little common sense here can spare the building unnecessary damage and save valuable time.

20. *D.* Ordinary plate glass demonstrates the characteristics described in choice D. Severe cuts or lacerations can result if caution is not exercised. The other choices given are not true.

21. *A.* Choice B is incorrect because the ridgeline of the two sections of the home should be perpendicular to one another, not linear. Choice C is incorrect because of the chimney placement. Choice D is incorrect because the ridgeline should not extend as far over into the other part of the home as is shown.

22. *B.* Since heat rises, a fire burns significantly faster in a vertical or uphill direction than laterally. Choice C is incorrect because higher humidity has a dampening effect on fire. Dryer climates are more conducive to fire. Choice D is incorrect because it can mean any number of situations that can actually inhibit fire progression. One such example of rough terrain is a rocky prominence. Even though it is rough terrain, less combustible fuel is present, and therefore a fire's growth or intensity is constrained.

23. *A.* Pentane has a lower flash point range of -129.8° Fahrenheit. Since it has the lowest flash point of the chemicals listed in the question, it would present the greatest hazard with regard to accidental ignition. If the chart had not taken into account the lower flash point of some chemi-

cals. Methyl vinyl ether would then be considered the most hazardous with regard to potential flammability.

24. **C.** Flash point variance can be determined by simply subtracting the lower flash point from the upper flash point of a given chemical. Therefore, pentane's variance is -40 − (-129.8) = -40 + 129.8 = 89.8°.
Pentanediol's variance is 275 − (-15.6) = 275 + 15.6 = 290.6°.
Monochloromethane's variance is 32 − (-97.7) = 32 + 97.7 = 129.7°.
Acetate's lower flash point was not given in the chart; therefore, a flash point variance cannot be determined. According to what has been figured, pentanediol's flash point variance is widest of the chemicals given, followed by monochloromethane and then pentane.

25. **D.** Acetate possesses a red classification and has a flash point greater than 182° F. Choice A has the same kind of classification but has an upper flash point of only 182° F. The question asks for a flash point that is greater than 182° F, not equal to it or less. Choices B and C do not possess a red classification and are therefore incorrect on that count alone.

26. **B.** Ethylbutylamine represents the highest upper flash point and monochloromethane represents the lowest upper flash point of those chemicals that are classified under yellow. (Hexanone has a midrange flash point of 57.2° F). This represents an upper flash point variance of only 32 degrees, which is fairly narrow. Choice A is incorrect because dihexylamine, which is classified under blue has the lowest flash point. Choice C is incorrect because the chemicals that fall under a red classification are sulfuric acid, pentanediol, and acetate, which have upper flash points of 182° F, 275° F, and 190° F, respectively. Flash points of this range are considered to be fairly stable. The lower a flash point is, the more dangerous that substance becomes with respect to accidental ignition. Choice D is incorrect as well because the chemicals that are classified under red and yellow (six chemicals) are more prevalent than green and yellow (five chemicals).

27. **B.** Rapid blood loss can kill someone in less than three minutes. It should be given priority over other first aid. The second most important step is to restore breathing because a victim can suffer significant brain damage if starved of oxygen for longer than 5 minutes. Although an injury may not be extremely severe, complications from shock can be life-threatening as well. This condition should be treated next. Choice D is one of the last things anyone has to worry about when it comes to stabilizing an accident victim.

28. **D.** Safety is paramount in firefighting. Anything contrary to that concept has no business within a fire department. It is never suggested to second-guess a commanding officer when the prospect of life and property are not endangered. An analysis of the event by a higher officer can be made at a later time. However, if it is ascertained by a firefighter that the orders issued may, in fact, endanger the lives of his co-workers or the public, it is within his or her rights to refuse compliance. It is better to risk the remote possibility of being reprimanded than it is to attend the funeral of a co-worker. Choice B is incorrect because the first thing that should be done is communicate to the commanding officer the reason for concern and only then refuse to comply if the original orders are insisted upon. At this point, it would be proper to have a higher officer intercede and make a decision on the matter. Choice C is not an appropriate way of dealing with such a matter.

29. **A.** Of the four choices given, only choice A warrants dismissal. Not only does a firefighter pose a threat to him- or herself under the influence of drugs, but he or she can directly endanger the lives of others. Drugs can induce complacency toward safety, ineptitude with regard to firefighting practices or just plain incompetence. None of these can be tolerated. The other choices given can be handled either by oral admonishment or an official reprimand.

30. D. Judging by the angles involved, the top floor would offer the worst vantage point for hose-stream penetration from a ground fixed nozzle. Water would immediately glance off the ceiling and fall close to the window opening. Hose stream penetration is minimized along with a diminished ability to extinguish a fire at that level. As the height is decreased, the vantage point offered is improved. The ground floor ultimately offers the best angle.

31. B. Any time heavy amounts of water are applied to an upper story, it is effective most of the time in diminishing a fire on the floor directly below it. It thereby lessens the fire's intensity and allows for smaller lines to be brought in on the floor that is burning to finish the job. The end may justify the means; however, a significant amount of water damage will occur to the building when this technique is implemented. Water will run down between walls, floor boards or other vertical arteries. The remaining choices are incorrect. This is particularly true of Choice D. You have to ask yourself, why would firefighters use an ineffective technique to combat structural fires in the first place?

32. D. As much as it seems strange, steel is not very fire resistant. If it is directly exposed to temperatures exceeding 1000° F, it can expand considerably and lose its strength. The approved coverings serve as insulating barriers. If the fire is not too intense, the protected steel support is less likely to be affected. Choices A, B, and C are false on their own merit.

33. A. A charged hose line is heavy and fairly stiff. If the option exists for reserve line to be looped above the fire floor, it is wise to do so. In this respect, less effort is required to stretch hose line on the affected floor. Choices B, C, and D ought to be taken into consideration; however, Choice A is the best reason to justify such a practice.

34. C. Water supply limitations are the least of a firefighter's worries when it come to fighting a fire on board a ship. Salt water may be suctioned into the ship by a pump to provide the necessary source of water. Choices A, B, and D are major concerns when fighting ship fires. It may be added that proper ventilation is sometimes very difficult to achieve because of the problem of limited access to various quarters in the ship.

35. C. Most smoke detectors operate by battery power. Considering their location within the home, it is somewhat inconvenient to replace the batteries. In addition, homeowners have a complacent attitude toward fire safety and something of this nature is easily forgotten or ignored. Choice A would affect a smoke detector's performance; however, it has been shown that this is not a valid concern. Choice B is false. Earlier models of smoke detectors may not have been as sensitive to heat or smoke as newer designs but, nonetheless, they were functional. Choice D does account for a significant percentage of fire detection failures; however, battery maintenance problems are far more prevalent.

36. C. If a ladder truck is placed below or downhill from the structure involved, two advantages are gained. The rung tilt is significantly reduced, giving the ladder truck some additional stability. If a ladder's reach is not overextended, an apparatus's center of gravity will remain within the safe limitations established by the manufacturer. Stationing an aerial ladder uphill from an involved structure will have just the opposite effect. Choices A and B do not afford any advantage for aerial ladder placement. Slope is a more important factor to consider rather than any one particular direction.

37. C. Choices A and B are incorrect because theft is a serious infraction of firefighting rules and regulations and warrants dismissal from the department. Something of this nature cannot be ignored. However, it would be foolhardy to do what Choice D implies. Can you imagine the repercussions of accusing a co-worker of theft, when, in fact, he or .she may have been collecting evidence as a part of an arson investigation. Choice C would be the most prudent action to take.

38. B. The location of a fire is considered most important because it can directly affect life hazard and extension probability. For instance, if a fire, regardless of its size or type, was to take place on the roof of a multilevel structure, the threat posed to occupants is less, and the probability that the fire will spread elsewhere is minimized. However, if the fire were located in the basement or lower floor of the same building, the threat or hazard posed to residents would be significantly higher. At this point, the size of fire, type of fire, and extension probability are all factors that warrant a greater degree of concern for firefighters.

39. D. Choice D represents the correct relationship. As water volume decreases within the tank, there is a proportional decrease in pressure, consequently slowing flow velocity at the discharge port. Choices A and C essentially represent the same concept, only in reverse order. Even if you have only partial understanding of flow dynamics, both of these alternatives can be immediately eliminated because there cannot be two correct answers within a question. Choice B is incorrect as stated because the last few gallons of water drained from the storage tank would exhibit low flow velocity and consequently take the longest time to discharge.

40. C. Fires pose more of a threat between midnight and 8 a.m. because sleeping people are slow to wake and to realize and react to fire related emergencies. In fact, carbon monoxide poisoning from a fire can suffocate a sleeping individual before the person has a chance to become cognizant of any danger. Additionally, these nighttime hours can present a visibility problem because of darkness. Without the aid of some illumination, evacuation of residents can be hazardous, and the difficulty for those conducting search and rescue operations compounded.

41. D. Choices A, B, and C are all reasons that tend to complicate school fires.

42. D. $(F° - 32)(5/9) = $ Celsius

$(99 - 32)(5/9) = X$

$67 \times 5/9 = X;\ X = \dfrac{335}{9} = 37.2$ degrees Celsius

43. C. A milliliter, as the name implies, means $1/1000$ of a liter. Another way to look at it is that there are 1000 milliliters in 1 liter. Therefore

12.38 liters $\times \dfrac{1000\ ml}{1\ liter} = 12{,}380$ ml

44. C. 96 kilometers can be converted to miles by using the

$\dfrac{96 \times 1}{1.6} = 60$ miles

45. A. Since we know that 5.735 inches represents 97.5% of the gear's pitch (100% − 2.5% = 97.5%), we can set up the following proportion to determine the actual pitch of the gear.

$\dfrac{97.5}{100} = \dfrac{5.735}{X}\ ;\ 97.5\ X = 573.5$

$X = 5.88$ inches (rounded).

ANSWERS TO PRACTICE EXAM 4

46. **A.** The square footage of a rectangle is found by multiplying its length by its width. In this case,

    47.5 ft × 13.2 ft = 627 square feet.

47. **B.** The volume of a cylinder is equal to $\pi \times R^2 \times$ Height. Since we are calculating the volume of acetyl chloride in the tank, its level (20.7 feet) will serve as the height of the cylinder. The radius is equal to half the diameter, or in this case, 18.5 ft × .5 = 9.25 feet. The volume of acetyl chloride = 3.1416 × 9.25^2 ft × 20.7 ft = 5564.2252 cubic feet. Therefore, to figure the number of gallons of acetyl chloride involved, we multiply this volume by 7.48 gallons.

    $$5564.2252 \text{ cubic ft} \times \frac{7.48 \text{ gal}}{1 \text{ cubic ft}} = 41620.404 \text{ gallons}$$

48. **C.** The area of the rectangle is length × width or 19.8375 meters × 15 meters, which is equal to 297.5625 square meters. Since we know that both areas are equal in size and that the area of a square is the length of one side squared, we can set up the following equation to solve the question:

    $$X^2 = 297.5625$$

    Therefore,
    $$X = \sqrt{1297.5625} = 17.25 \text{ meters}$$

49. **D.** The equation to determine the area of a triangle is (½) × base × height. With the 4:3 ratio existing between the base and its height, we can write the following equation to determine the length of side BC.

    $$(½)(4X)(3X) = 157.6 \text{ square meters}$$

    $$\frac{12 X^2}{2} = 157.6$$

    $$12 X^2 = 315.2$$

    $$X^2 = 26.267; \quad X = 5.125$$

    Therefore, the base AB = 4 × 5.125 = 20.5 meters and the height BC = 3 × 5.125 = 15.375 meters.

50. **B.** The best way to work this kind of problem is to use the equation below. The amount of 100% antifreeze (1.00 X) + the amount of 45% antifreeze solution .45 (14 − X) = amount of pure antifreeze in 75% solution (.75 × 14).

    1 X + 6.3 − .45 X = 10.5
    1 X − .45 X + 6.3 = 10.5
    .55 X = 4.2
    X = 7.636 or 7.64 quarts of 100% antifreeze

    To verify that this is indeed the appropriate amount, plug 7.64 into the equation designed for this problem and see if it holds true:

COMPLETE FIRE-FIGHTERS EXAM PREPARATION BOOK

$$1 . x\ 7.64 + .45\ (14 - 7.64) = 10.5$$
$$7.64 + 2.862 = 10.5$$
$$10.5 = 10.5$$

Therefore, 7.64 quarts of pure antifreeze is the correct amount to use to get a 75% solution of antifreeze under the conditions given.

51. *B.* First we need to convert 132,000 square feet into square meters:

$$132,000\ ft^2 \times \frac{1\ m^2}{10.764\ ft^2}2 = 12,263.1\ m^2$$

If we know that a Class A fire extinguisher can protect a maximum area of 557 m(2), we simply divide 557 m² into 12,263.1 m² to determine the required number of extinguishers.

$$\frac{12,263.1\ m^2}{557\ m^2} = 22.02$$

Since we cannot have 0.02 of a fire extinguisher, we must round up to the next whole number, which is 23. Choices A and C are wrong for this very reason.

52. *A.* The volume of a cylinder is equal to $\pi \times R^2 \times$ Height. To find the radius we simply multiply diameter by .5.

$$22\ inches \times .5 = 11\ inch\ radius$$
$$\pi \times 11^2\ inches \times 66\ inches = Volume$$
$$\pi \times 121\ inches \times 66\ inches = Volume$$
$$Volume = 25,088.82\ inches^3$$

Since there are 231 inches³/gallon, we divide 231 inches³ into the volume found to determine the number of gallons this container can hold. Therefore,

$$\frac{25,088.82\ inches^3}{231\ inches^3} = 108.61\ gallons$$

53. *C.* Since pulley A rotates clockwise and pulley D rotates clockwise as well, we can think of this question as being a direct compound proportion rather than an inverse one. An inverse compound proportion would involve directional changes of pulleys, accomplished by crossing the belt. This concept is further elaborated on in the mechanical principles section of this book. Therefore, the ratio 13/9 and 16/6 should be multiplied together. The resulting product demonstrated to be equal to X/54 rpm constitutes the proportion

$$\frac{13}{9} \times \frac{16}{6} = \frac{X\ rpm}{54}$$

$$\frac{208}{54} = \frac{X\ rpm}{54\ rpm}$$

$$X = 208\ rpm$$

54. A. This question involves a direct compound proportion to solve for the weight of the larger bar. The fact that the larger the bar is, the heavier it becomes is considered to be a direct proportion. Therefore,

$$\frac{4 \text{ ft.} \times 2 \text{ in.} \times 2 \text{ in.}}{7 \text{ ft.} \times 3 \text{ in.} \times 4 \text{ in.}} = \frac{12.3 \text{ pounds}}{X \text{ pounds}}$$

$$\frac{192}{1008} = \frac{12.3 \text{ pounds}}{X \text{ pounds}}$$

$$192 X = 1008 \times 12.3; \quad X = 64.57 \text{ pounds}$$

To verify that this is the correct answer, calculate the volume of each bar in cubic inches. If we know that the smaller bar has a volume of 48" x 2" x 2" or 192 inches^3 and weighs 12.3 pounds, we can simply divide 192 into 12.3 to determine the weight per inch^3. In this case, it equals .06406 pounds. Then multiply .06406 pounds to the volume found for the larger bar (84" x 3" x 4" = 1008 inches^3) which equals 64.57. Thus, 64.57 = 64.57 and our answer is correct.

55. C. Choice C best describes what the article said. Choices A and D are too specific or focused and fail to outline in general what the article was about. Choice B is an incorrect statement according to the reading.

56. B. Carbon monoxide is chemically known as CO. CO_2 is carbon dioxide, O_2 is oxygen, and COC_{12} is phosgene.

57. D. 1%, or .01 in decimal form. Choice B represents adequate concentration of CO to displace approximately 50% of the oxygen present in a person's bloodstream.

58. C. This may seem to be a confusing question but all the options given describe the various sizes of four different individuals. Since no physical ailments of the kind that affect CO asphyxiation were mentioned, we are left only with physical statures to determine who is more susceptible. Choice C is the shortest and most petite individual described, therefore the most susceptible to CO poisoning.

59. A. We can assume that since a barbecue grill is being used to heat a room during a cold spell, we can consider that the room is probably sealed as well as possible to prevent what little heat there is from escaping. In a closed environment such as this, the conditions are ideal for CO buildup. Without any ventilation it would only be a short time before concentrations of carbon monoxide would accumulate. Choices B and C are situations where CO is present, but because both are outside, there is ample ventilation to prevent CO asphyxiation from occurring. Propane, on the other hand, is a fuel used for indoor heating and cooking. The reason it is safe to use in a closed environment is that it burns efficiently and completely. CO is not a by-product of propane combustion.

60. C. An individual subjected to the first stages of CO asphyxiation initially experience an increased pulse and respiration. The remainder of symptoms given are associated with CO poisoning.

61. B. The acronym for Personal Alert Safety System is PASS or, in other words, the initial letters of the words in the name of the device spell PASS.

62. D. The last statement in the reading implies that even though the technology to improve firefighter safety has been developed, not all firefighters take advantage of the benefit. This may hold true for volunteer fire departments that lack the necessary funds to purchase such systems or for a host of other reasons not detailed. Choice A is not an implication; rather, it is taken directly out of the last statement in the reading. It is a fact. Choice C was mentioned in the article; however, it fails to provide what the question wants. Choice B was never clarified in the passage.

63. C. This is kind of a trick question. By switching PASS to either the ON or ARMed position, it can be determined if the batteries in the unit are sufficiently charged. The "ON" setting, if you recall, emits a continuous alarm and ARMed emits a short alarm to indicate that the device is functioning properly.

64. D. All options given would be reasons that make it advantageous for firefighters to wear a Personal Alert Safety System.

65. A. Once PASS is armed, it requires a firefighter not to move for a period of only 20 to 30 seconds before an alarm is sounded.

66. B. The article essentially describes what is necessary in implementing a home fire safety program. The benefits to be gained were made fairly obvious by the article. Even if the reader did not recognize the benefits gained by following such a safety program, the last statement in the article makes direct mention of the advantage involved. Choices C and D were both discussed in the passage; however, neither completely summarizes the intent of the article.

67. C. EDITH stands for Exit Drills In The Home, according to the article.

68. D. The National Fire Protection Association is the founder of both home fire safety programs discussed.

69. A. A predetermined meeting place must be an outdoor location that is a safe distance from the house. Having residents congregate at one particular location in the home during a fire poses extreme risk to everyone involved. Choices B, C, and D are correct measures taken when implementing the home fire safety programs discussed.

70. D. Master bedrooms in split-level homes are upstairs and high enough off the ground to cause injury to persons attempting evacuation through the bedroom windows. When this kind of situation exists, it is recommended that a ladder of appropriate size be installed to provide residents a safer avenue of escape. Choice A would be correct if it were not for the fact that the room described was not upstairs. Choices B and C are means of improving safety in other kinds of homes; however, neither was discussed within the reading.

71. D. The underlying intent of the article was to make the flow of electrical energy easy to understand. Choice A is wrong because it was not differences between water flow and electrical flow that were elaborated on; rather, it was the similarity shown between the two that was discussed. Choices B and C were concerns within the reading, but do not cover the intent of the article.

72. A. Electrical energy transfer is accomplished by electron migration at the atomic level.

73. C. Wood, especially when it is dry, is a poor conductor, which makes it a good insulator. Metals transfer electrical energy better.

74. B. Remember the correlation mentioned in the reading. If electrical flow is marginal or poor, the

resistance in ohms is greater. Choices A and D mean basically the same thing. Choice C is a factor more applicable to water flow than electricity.

75. *A.* Amperes is a quantitative measure of electrical energy flow or current. Voltage represents electrical pressure or potential. Ohms represents electrical resistance.

76. *C.* Ohm's law specifies that electrical pressure in volts is equal to the resistance in ohms multiplied by the current in amperes ($V = O \times A$). To further demonstrate that choice C is correct, let's apply some arbitrary numbers to the formula given. If $V = 10$ and $O = 2$, we know that $A = 5$. However, if $O = 5$, we know that $A = 2$. Therefore, if V remains the same and O increases in value, we know that A must proportionally decrease. If A increases in value, O must proportionally decrease. If you try the other alternatives, you will find they are not compatible with to Ohms law.

77. *B.* Any kind of market depends on supply and demand fundamentals. If a material is hard to come by or is relatively rare, it can command a higher price than that material which is more common. Silver and gold are considered precious metals because of their scarcity. Consequently, they are both expensive commodities when compared to copper. Copper is a more readily available material. The remaining choices given are false.

78. *B.* The intersection of Duff and Maple falls under the jurisdiction of Fire Station #9.

79. *C.* The triangular block described in the question falls under the jurisdiction of Fire Station #13.

80. *A.* The northwest corner of the block bounded by 6th and 8th Avenues, Lilac Lane, and Parker Place is the most central point in Fire Station #5's jurisdiction. Choice B could have been correct had it referred to the southeast corner instead of the southwest corner. The other two choices were too far away from the central area of the jurisdiction to warrant consideration.

81. *D.* Fire Station #17's jurisdiction encompasses the smallest area. Additionally, Long Lake constitutes a good portion of the jurisdiction itself. Consequently, there are probably fewer residences in that area due to lack of space.

82. *B.* The intersection of Miller Road and Wright Avenue.

83. *C.* Northeast. If the question asked "where is the intersection of 14th Avenue and Parker Place in relation to 8th Avenue and Jenson," choice D would have been the correct answer.

84. *A.* Northwest

85. *A.* Fire Stations #5's jurisdiction.

86. *D.* Go east two blocks from the station, then two blocks north.

87. *B.* South. The direction given would describe the vehicle heading south on Blackmore.

88. *D.* Gears 3,5, and 7 turn counterclockwise. Gear 2 turns clockwise.

89. *C.* The length of the handle that extends perpendicularly from the windlass's drum is the key to the amount of leverage exerted on the line. The longer the handle, the larger the amount of force that can be exerted on the smaller wheel or in this case, the drum. Choices A, B, and D have little if any bearing on leverage gains.

90. *D.* Since we are dealing with a wheel and axle, we can use the following formula to deter-

mine the answer: Force x Circumference of large wheel = Resistance x Circumference of small wheel. We already know that the force is equal to 125 pounds. The circumference of the larger wheel, which represents the handle's turning circumference, is our unknown. Resistance is the amount of tension on the line, which is equal to 400 lbs. The circumference of the smaller wheel or, in other words, the windlass drum is equal to π x 6"(diameter) = 18.85". If we plug this information into the formula, we can solve for our unknown by using simple algebra:

$$125 \text{ lbs} \times X = 400 \text{ lbs} \times 18.85 \text{ inches}$$

Therefore,

$$X = \frac{400 \times 18.85}{125 \text{ lbs}} \; ; \; X = 60.32 \text{ inches}$$

This is not our answer, however. This number is the circumference of the larger wheel. We are after the diameter instead, to reflect the necessary handle length for the circumstances given. Therefore, setting up the problem as below will give us the correct answer:

$$\pi \times X \text{ diameter} = 60.32 \text{ inch circumference}$$

$$X = \frac{60.32 \text{ inches}}{3.1416} \quad X = 19.2 \text{ inch handle is necessary}$$

91. **B.** If we already know that 125 pounds of effort can achieve 400 pounds of line tension or pull, we can simply divide resistance by effort to determine the mechanical advantage.

$$\frac{400 \text{ lbs. resistance}}{125 \text{ lbs. effort}} = 3.2 : 1$$

92. **B.** The formula that applies to inclined planes is: effort x length of inclined plane = resistance x height. Effort is what we are trying to solve for. The resistance is the weight of the drum (460 lbs). According to the diagram, the height of the inclined plane used is 4.25 feet. The length of the inclined plane warrants extra consideration. We can look at the inclined plane as a right triangle. According to the Pythagorean theorem, when we square the base and height of a right triangle and add them together, we can determine the hypotenuse (i.e., length of the plane itself) by taking that sum and determining its square root:

$$9.25^2 + 4.25^2 = 85.5625 + 18.0625 = 103.625$$

$$\sqrt{103.625} = 10.18 \text{ feet.}$$

Now that we know the length of the plank or plane itself, we can plug this information into the formula for inclined planes and solve for our unknown.

$$\text{Effort } (X) \times 10.18 \text{ feet} = 460 \text{ lbs} \times 4.25 \text{ feet}$$

$$X = \frac{460 \text{ lbs} \times 4.25 \text{ feet}}{10.18 \text{ feet}}$$

ANSWERS TO PRACTICE EXAM 4

$X = 192$ pounds, the total effort required for the task. However, since two firefighters are involved in rolling the drum up the inclined plane, we can assume that each firefighter need only contribute half the total effort required. Therefore, we get 192 divided by 2 = 96 pounds of effort per individual.

93. **C.** The question does not concern itself specifically with mechanical advantage. If that were the case, it would account only for the benefit gained by using the inclined plane. If this had been what the question was after, Choice A would have been the correct answer. However, the questions was worded "all things considered," which would include the help rendered by the other firefighter. Since it was determined that each firefighter needed to exert only 96 pounds of effort, we can simply divide this into resistance to calculate lift advantage.

$$\frac{460 \text{ pounds weight}}{96 \text{ pounds effort}} = 4.8 : 1$$

94. **A.** Since this pulley demonstrates a 2:1 mechanical advantage (because of the two ropes in the pulley arrangement), we simply divide the resistance or weight being lifted by 2. Therefore, 375 lbs. divided by 2 = 187.5 lbs. of effort required.

95. **A.** To determine the height that the weight will lift, we need to use the following formula.

    Length of pull = Lift x Mechanical advantage

    Therefore, 15 feet = (X) x 2

    $$\frac{15 \text{ feet}}{2} = X; \quad X = 7.5 \text{ feet}$$

96. **D.** Choice A is incorrect because wheels 3 and 4 rotate at the same rpm since they are keyed to the same shaft. Therefore, wheel 4 turns at the same speed that wheel 2 does. Choice B is incorrect because wheel 5 turns in a counterclockwise direction. This is the result of twisting the beltdrive between wheels 2 and 3. Choice C is incorrect, because wheel 5 turns at a greater rpm than wheel 2. Wheel 4 is directly responsible for wheel 5's speed enhancement.

97. **B.** The first thing we need to do is convert tons into pounds so that everything is figured in one unit of measure—in this case, pounds. Since there are 2000 pounds/ton

$$2\frac{1}{2} \text{ tons} \times \frac{2000 \text{ lbs}}{1 \text{ ton}} = 5000 \text{ lbs}$$

To determine the mechanical advantage of this jack screw, we need to calculate the handle/lever's turning circumference. Since the handle length represents only the radius of a circle, the circumference is found by multiplying the radius by 2 and then multiplying by π (3.1416). Therefore, 19 inches x 2 x π = 119.38 inch circumference. Now, mechanical advantage can be determined by dividing the screw pitch into the circumference found:

$$\frac{119.38 \text{ inches}}{\frac{1}{8} \text{ inch pitch}} \text{, which is the same as saying, } \frac{1}{8} \text{ inch pitch}$$

119.38 x 8 = 955.04

$$\frac{\text{Resistance}}{X \text{ lbs. of effort required}} = \text{Mechanical advantage}$$

Therefore,

$$\frac{5000 \text{ lbs}}{X \text{ lbs of effort}} = 955.04 \times X$$

$$X \text{ lbs of effort} = \frac{5000 \text{ lbs}}{955.04} = 5.24 \text{ lbs of effort is required}$$

98. **D.** Since the drive gear has twice the number of teeth as the driven gear, there is a 50% reduction of torque with a corresponding 50% increase in speed.

99. **C.** To understand this problem better, let's approach it in a quantitative sense. Let's assume for the moment that gear A (the driver gear) is rotating at 1000 rpm. Since we are told that gear A is half the size of gear B, we can figure that rpm are reduced 50% or in other words, gear B will turn at 500 rpm. Gears B and C will turn at the same rate because both gears are keyed to the same shaft. However, gear C is half the size of gear D, so another 50% reduction of speed occurs. In the end, gear D rotates at 250 rpm or ¼ the drive gear's speed. Remember, what is sacrificed in speed is gained in torque. Therefore, four times the amount of torque is realized at the expense of speed.

100. **A.** This question involves several mathematical steps. If we are told that we want to raise the level of the third cylinder from 4.5 liters to 9.75 liters, that represents an added volume of 5.25 liters (9.75 liters − 4.5 liters = 5.25). Since the tank has a 13-liter capacity, this reflects a .4038 or 40.38% increase in volume. Now that we have the added volume figured as a percentage, we need to calculate comparable volume increases in the other two cylinders. If cylinder 2 were to increase its volume by 40.38%, that would account for 2.826 liters of water. If cylinder 1 also has a 40.38% increase in volume, that accounts for another 2.019 liters. To reflect the total volume of water needed to increase the levels in each tank, we simply add the volume differences together to arrive at the final answer. 5.25 liters + 2.826 liters + 2.019 liters = 10.095, so 10.1 liters of water are required.

TEST RATINGS ARE AS FOLLOWS

95–100 correct	EXCELLENT
87–94 correct	VERY GOOD
81–86 correct	GOOD
75–80 correct	FAIR
74 or below correct	UNSATISFACTORY

Go back to the questions you missed and determine if the question was just misinterpreted for one reason or another, or if it reflects a particular weakness in subject matter. If it is a matter of misinterpretation, try reading the question more slowly while paying close attention to key words such as *not, least, except, without,* etc. If, on the other hand, you determine a weakness in a particular area, don't despair. That is what this book is for; to identify any area of weakness before you take the actual exam. Reread the material on the area of concern in this book. If you still feel a need for supplemental material, your local library is an excellent source.

PHYSICAL FITNESS

Considering the potential physical demands that firefighting may entail, it is not hard to understand why fire departments require employees to be in top physical shape. In-station tasks require little exertion; however, when firefighters are called to active duty, the switch from a sedentary pace to extreme physical exertion is stressful. This is particularly true for someone who is out of shape. Some of the demands of the job will involve running, climbing, jumping, twisting, pulling, and lifting heavy tools and equipment, all the while being constrained by protective gear. This is also compounded by the fact that work is performed under temperature extremes, poor air quality and/or visibility, and water exposure. A firefighter must be physically able to respond to these conditions while always being careful of his or her own safety. This is the primary reason fire departments place equal emphasis on the physical fitness exam and the written exam.

Physical fitness exams are quite varied. What one department considers suitable may be considered inadequate by another department. To further illustrate this point, two physical fitness exams are outlined below. Both tests measure strength, stamina, and flexibility, but each determines these capabilities in a completely different manner.

FIRE DEPARTMENT A

1. Bench Press: Candidate lies on his or her back and is required to press a 95-pound barbell directly overhead with arms fully extended. Five consecutive presses are required to pass.
2. Ventilation Fan Lift: Candidate must lift a ventilator fan that weighs approximately 60 pounds off the floor and hang it on an overhead bar using the hooks on the back of the unit. After the unit is hung, it must immediately be put back on the floor in its original position. PASS/FAIL.
3. Arm Curls: Candidate stands with back against a wall and is expected to arm curl an 85-pound barbell. The procedure starts with the candidate's arms fully extended along his or her side and, without arching the back or swinging the weight, he or she must bring the weight bar upward to the shoulders. Five consecutive curls are required to pass.
4. Hose Carry: Candidates conduct this exercise while wearing an air mask that weighs approximately 25 pounds. This is a timed event. When a candidate is instructed to begin, he or she has to remove an 80-pound hose pack from where it is stored and will be required to run up a flight of stairs to the seventh floor of a training tower. Any time in excess of 1 minute, 30 seconds is deemed a failure.
5. Equipment Hoist: Candidate must hoist an 80 pound weight to the seventh floor of the training tower. He or she has the option of using either a hand-over-hand pull or a two-handed pull. This event is timed as well; anything over 35 seconds is considered a failure.

FIRE DEPARTMENT B

1. Aerial Ladder Ascent: Candidate, wearing a firefighter's coat, mask, and air tank, climbs a 75-foot aerial ladder. This is not a timed event; however, it is important that a slow and steady pace be

maintained without slipping or stopping for rest. Once at the top, the candidate must immediately descend to ground level. (Note: Candidate is secured to a safety line for this event.)
2. Balance/steadiness: Candidate, wearing the same equipment used during the aerial ladder ascent exercise, walks on the rungs of a 25-foot ladder suspended horizontally three feet off the ground. Falling off the ladder or stepping in between the rungs results in failure.
3. Shoulder Hose Carry: Candidate lifts an 85-pound hose pack off the ground and drapes it over his or her shoulders. Candidate must then run a 50-yard figure-eight course. This must all be accomplished within 35 seconds in order to pass.
4. Charged Hose Drag: A 150-foot length of 2½ inch charged hose is laid perpendicular to a pumper truck. The candidate must take the nozzle end of the charged line and bring it back to the pumper. This must be accomplished in less than one minute.
5. One Mile Run: Candidate must run one mile in less than seven minutes to pass.

Failure of any one of these exercises constitutes disqualification.

OTHER TYPES OF TESTS SEEN ON PAST FITNESS EXAMS

Other features seen on past physical fitness examinations include the following:

1. Tunnel Maze: Candidate must crawl through a tunnel or some other form of labyrinth within a specified period of time. This test simulates firefighters crawling through attic spaces or crawl spaces beneath homes.
2. Ladder Raise: Candidate must lift a ladder that is 25 feet in length and weighs approximately 75 pounds, off of the ground and raise it against some form of backdrop. This is a pass/fail test.
3. Wall Scaling: Candidate must vault him- or herself over a four- to six-foot-high wall without the assistance of a ladder or bench. This is a pass/fail test.
4. Simulated Victim Carry: Candidate must carry or drag a 160 pound dummy approximately 75 feet within one minute to pass.
5. Sit-ups or Pull-ups: Candidate has exactly one minute to accomplish as many sit-ups or pull-ups as possible. Anything below 20 is considered a failure.
6. Hose Coupling: Candidates are given brief instructions on how to connect and disconnect a hoseline coupling. Each candidate is then expected to disconnect a hose coupling and reconnect it to another hose coupling within one minute to pass.
7. Forcible Entry Simulation: This may require the candidate to force a door open or use a fireman's axe to breach a barrier (flooring, partitions, etc.).

A few of these exercises also allow examiners to scrutinize an applicant's psychological fitness. For instance, does the candidate have a fear of heights or exhibit some degree of claustrophobia when wearing firefighting equipment or working in confined spaces? These factors are taken into consideration to determine a test applicant's true potential as an effective firefighter.

Bear in mind that all factors will be taken into consideration when determining a test applicant's eligibility for hiring. That is why it is imperative not to take anything for granted, regardless of the situation you are in. Practice these events in trial runs as best you can before the actual exam. It is better to learn of potential weaknesses beforehand, rather than fall short during a timed event and fail the test altogether. You have come too far to let this job opportunity slip away simply because of physical unpreparedness. Approach this half of the exam in the same manner as you did the written test. By practicing the workout schedules suggested below, not only will you be in better shape, but you will be able to approach the physical fitness exam with more confidence and a sense of ease.

Before charging into any fitness workout, however, it is suggested that you visit your family doctor and

PHYSICAL FITNESS

get a complete medical evaluation. This will be a precondition to your employment with the fire department in order to determine if any disease or physical condition may impair your ability to perform. If a condition may not now impair potential work performance but may manifest itself at a later time, it may disqualify a test applicant from further consideration for employment. Each fire department has its own basic guidelines. Some physical conditions are tolerated, while others may cause complete rejection. Contact the personnel office of the fire department you intend to apply to and procure a copy of a medical evaluation form or information on fitness requirements.

Prior to actually taking the physical fitness exam, you will be given a medical questionnaire to complete. The questions will concern everything from allergies to urinary conditions. m is provides examiners with general information to make an educated decision as to whether such an exam poses a health risk to a test candidate. Of course, everything marked on this questionnaire will later be verified by a fire department medical examination.

If, in fact, you have a condition that warrants rejection from employment consideration, consult your doctor. He or she may be able to prescribe treatment through a change of diet, specialized workout, medication, or surgery to correct the problem. Inform your physician of your intentions concerning any kind of physical workout. No two people are the same; thus, workout schedules vary. The guidelines provided in this book are just that, guidelines. Your doctor is better able to tailor a training program that will benefit you. Keep your doctor's advice in mind while you prepare for the exam.

Regardless of the variance seen in physical fitness exams, it can be generally surmised that three kinds of physical attributes are being scrutinized during these exams: flexibility, cardiovascular fitness, and muscular strength and Endurance. Each of these areas use different groups of muscles and the exercises suggested below for each will improve them.

However, it is important to realize that prior to any exercise there are preliminaries that can help to prevent injury to yourself.

RELAXATION AND WARM-UP ROUTINE

The first thing to do before starting any rigorous exercise is a relaxation and warm-up routine which involves the head roll, paced breathing, and shoulder shrugs.

For the head roll, you can be either standing, sitting, or kneeling. Allow your head to go limp and roll it around your neck two or three times in one direction and then two or three times in the other direction. It helps to close your eyes during this exercise to prevent any dizziness or loss of balance. Try to conduct this exercise slowly and smoothly.

Paced breathing involves lying on your back and placing your hands on your stomach. Concentrate on your breathing by paying close attention to how far your chest rises during each inhalation. Breathe evenly and slowly and relax for about one minute during this exercise.

The third relaxation exercise is shoulder shrugs. Again, lie on your back. Simply pull your shoulders upward and maintain that position for a few seconds before allowing your shoulders to return slowly to their original position. Try to coordinate your breathing so that you inhale while pulling your shoulders up and exhale when your shoulders drop. Perform this exercise for approximately one minute.

FLEXIBILITY EXERCISES

Stretching exercises are important, too, because they prepare tendons and ligaments for further stretching and increase the flow of fluid around various joints. The whole concept is based on smooth, even and slow motion. This kind of exercise is not intended to be conducted in fast or jerky movements.

The back stretch or swivel is the first flexibility exercise to do. Standing, and with your arms at your sides, try to lean as far forward as possible. Then lean as far backward as you can. Repeat these exercises at least four times in both directions. Now, to limber up your back for bending sideways, remain standing, turn your head to the right, and slide your right hand down the length of your right leg as far as possible. Do this exercise at least four times on the right side, then four times on the left side.

To stretch the quadriceps (thigh muscles), stand and lean against a wall using your left hand as support. Reach behind with your right hand and lift your right leg up so that you can grasp your toes. Slowly pull your heel closer to your buttocks until the thigh feels stretched. Maintain this position for approximately five seconds. Repeat this exercise four times with the right leg before doing the same with the left leg.

To stretch the calf muscles in your leg, remain standing within arm's length of the wall. While facing the wall, keep your feet flat on the ground (do not allow your heels to lift), and allow yourself to lean forward for a few seconds. Push off against the wall to return to the starting position. Repeat this exercise three or four times with each leg.

Now while sitting on the floor with your legs spread apart and the back of your legs flat on the floor (your knees should not lift), slide both hands as far down the leg as possible. Hold this position for a few seconds before sitting erect again. Repeat this exercise three to four times and then do the same for the other leg. This exercise stretches both back and hamstring muscles.

Remain in the sitting position and cross your legs, putting the soles of your feet together. Now, lean forward as far as possible and hold this position for a few seconds before sitting erect again. Repeat this exercise three or four times. This exercise stretches the groin muscles.

To stretch the hips, remain in the sitting position with your legs straight. Now, take your right leg and cross it over the left leg. Take the knee of the right leg and slowly bring it up to your chest. Hold that position for three to four seconds and then repeat this exercise twice more. Do the same with the left leg.

The final exercise involves stretching chest, shoulder, and back muscles. While kneeling, place your palms on the floor, then slowly slide both hands forward until your elbows touch the floor. Keep your head and back straight during this exercise. Return to your starting position and repeat this exercise three or four times.

Remember, the whole point of these exercises is to stretch various muscles. If you force a muscle to extend too far, pulling or tearing can occur which defeats the purpose of stretching. Stretch various muscles only to the point of mild sensation, hold for a few seconds, then relax. This has the effect of increasing flexibility and loosening muscles for other exercises.

The last preliminary needed before any exercise is a cardiorespiratory warmup. This simply involves conducting an exercise that is not too stressful, such as brisk walking or slow jogging for a few minutes. This allows the heart rate to increase gradually and prepares the heart for vigorous exercise. To prevent potential injury, a warmup routine should always be done before any stressful exercise.

CARDIOVASCULAR FITNESS

Cardiovascular fitness simply measures your heart and lung capacity. As both of these organs become more fit, your body's ability to transport oxygen to various cells improves. Another beneficial result is that the heart beats less but pumps with greater strength—or, in other words, works more efficiently. There is also a corresponding increase in the peripheral circulatory system, thereby making it easier for various cells to absorb oxygen.

The best way to achieve cardiovascular endurance is to employ what physiologists call aerobic exercise. This may come in the form of one of four events: running, swimming, bicycling, or walking. When any of these forms of exercise is conducted fairly rigorously for approximately 25 minutes three times a week, cardiovascular endurance will improve. The key point here is to exercise at a moderate intensity, nonstop for the full 25 minutes. Less makes the exercise much less useful. That is why sports such as baseball, tennis, or basketball do not suffice. These sports require tremendous energy output some of the time; however, there are breaks in between. To be effective, the exercise has to be conducted for 25 consecutive minutes, stopping only to check pulse.

To calibrate your progress using aerobics, physiologists have come up with a pulse-rated system. Pulse essentially takes into account the number of times your heart beats per minute. As your cardiovascular endurance improves, your heart beats less when subjected to stress. To measure pulse, simply apply one or two of your fingers (not your thumb) to the front of your neck next to the larynx and feel for the carotid artery. The pulse should be fairly obvious there. Be careful not to press too hard on this artery because unconsciousness may result, particularly after exercise. Count the number of times your heart beats within 10 seconds, and then multiply that number by six. This will provide you with an accurate assessment of your pulse. When performing a rigorous exercise for 25 minutes, stop after the first 10 minutes to take a brief pulse (10 seconds), and immediately resume the exercise.

Intermix the four events of running, swimming, bicycling, and walking in your training. This helps to alleviate boredom and perpetuates the desire to continue training. When your 25 minutes of exercise is completed, it is necessary to follow it with a cooldown period. Walk or jog slowly for 5 to 10 minutes. The general idea is to permit your body to return to its normal condition gradually. This cooldown can be followed by a few stretching exercises as well.

MUSCULAR STRENGTH AND ENDURANCE

Strength development can be accomplished by weight training and calisthenics. Both areas improve muscular endurance through repetitious movement but do so in different ways. Calisthenics essentially use exercises that employ your own body weight to serve as resistance. On the other hand, weight training involves lifting progressively heavier weights or resistance in the form of bar bells or variable resistance weight training equipment.

Calisthenics, like cardiovascular endurance exercises, need to be proceeded by relaxation, stretching, and warmup exercises. A daily routine of push-ups, sit-ups, pull-ups, leg raises, and squats should be conducted over a period of 15 to 25 minutes. Start out doing 15 repetitions of each exercise and then work your way up to accomplishing 30 repetitions. Don't expect this to occur overnight. Regularity is the key. Your persistence will reward you with greater strength within three to four weeks. Descriptions of each exercise are given below.

Push-ups
Lie on your abdomen on the floor and place your hands palm down beneath your chest. As you extend your arms and push off of the floor, be sure to keep your back and knees straight. Once your arms are fully extended, lower yourself to the floor slowly and repeat the exercise.

Sit-ups
Lie on your back on the floor and either have someone restrain your feet from lifting or place your feet beneath a sofa or other heavy object. Your knees should be straight and flat. With your hands locked behind your head, sit up and attempt to touch your knees with your elbows without lifting your knees. Do not try to force yourself to extend beyond what is comfortable. Stretch as far as possible and then return to the starting position to repeat.

Pull-ups
Use a chinning bar that is just a few inches higher than your highest reach when standing up and your arms are extended overhead. Using an overhand grip on the bar, raise yourself to the point where you bring your chin level with the bar. Try not to kick or swing while raising yourself. Lower yourself slowly to the starting position and repeat.

Leg Raises
Lie on the floor on your right side with your legs kept straight and in line with one another. Use your left arm to gain support. Lift your left leg as far as possible before returning to the starting position. Repeat this exercise a minimum of 15 times before changing sides and doing the same exercise with the other leg.

Squats
In the standing position, extend your arms forward and then squat until your thighs become parallel to the ground. Return to the standing position and repeat the exercise.

Weight training, when done correctly, significantly increases muscular strength and endurance. However, three things should always be kept in mind before starting any kind of weight training routine. Supervision by either a professional weight training assistant or someone to act as a safety person during your lifts is essential. This is particularly true while bench pressing barbells. The second consideration is always to begin light and progressively increase the weight you lift as you become stronger. Starting heavy is an open invitation to injuring muscle tissue instead of building it. The third consideration is to conduct a weight training routine only three times per week at the maximum. Keep the number of repetitions to only three sets of ten. Doing more will tend to increase bulk rather than strength. If the repetitions seem fairly easy initially, increase the weight load by 5 or 10 pounds at a time. Continue this progressive addition of weight as your strength improves. Below are exercises that concentrate on developing muscles needed most for firefighter fitness exams: chest, shoulder, arm, and back.

Bench Press
For safety considerations, it is better to use bench press equipment rather than free weights. Whichever is available, lie on your back and grip the bar with both hands at shoulder width. Begin with light weights, as mentioned earlier, and lift or press the bar directly perpendicular to the chest by extending the arms. Try not to lock your elbows when fully extended. Slowly lower the bar to your chest and repeat the exercise.

Arm Curls
While standing, preferably with your back to a wall, allow your arms to be fully extended downward. Grasp the barbell with an underhanded grip with both hands spaced shoulder width. Raise the barbell to your chest without allowing your elbows to move from your side. Lower the barbell to the starting position and repeat.

Half Squats
This is the same thing as done with squat calisthenics. The difference is that a barbell rests on the back of your neck while it is supported by both hands at shoulder width. As the weight is steadied on your shoulders, conduct squat repetitions as described under calisthenics.

Bent Over Rows:
Begin in the standing position with the weight bar on the floor directly in front of you. While keeping your legs straight, lean over the barbell in such a way that your back becomes parallel to the floor. With an overhand grip, grasp the weight bar with both hands spaced shoulder width and lift the weight to your chest. Try to keep your back straight (i.e., parallel to the floor) and your head up while attempting the lift. Return the weight to the floor and repeat the exercise.

As a final suggestion, try to improve your grip strength by either using a spring grip or (to better measure your forearm, hand, and finger strength), a Stoelting Grip Dynamometer. The one thing common to most firefighter physical fitness tests is grasping firehose, charged or uncharged, to pull, hoist, or carry it for various distances. Improving strength in this area gives you a definite advantage during the test.

THE INTERVIEW—THE FINAL PHASE OF SELECTION

Now that you have reached this point in the selection process, there are a few things to be aware of before an interview. You will be notified by mail of the time and place of the interview. Pay particular attention to the date and become familiar in advance with the location of the interview. One sure way to disqualify yourself from serious consideration is to show up late for the interview. There really are no excuses.

Appearance is also important. Most people are told not to judge others by outward appearance; however, interviewers gain a distinct impression from the manner in which a candidate dresses. If an applicant is not well groomed (e.g., unshaven, hair uncombed), interviewers have already perceived that candidate, without so much as asking one question, as being uncaring and somewhat sloppy. Even though the candidate may be the most hardworking and concerned individual among those being interviewed, he or she will, in all likelihood, be passed over for another with better appearance. First impressions are just as important as how you respond to questions asked by the interviewers. Therefore, be well groomed for the occasion and dress neatly. For men, this would entail a nice shirt (tie is optional), slacks, and a pair of dress shoes. For women, an attractive blouse, dress pants (or suit or skirt) and shoes or a conservative dress would be appropriate.

Another thing to avoid prior to or during an interview is smoking or chewing gum. Habits like these can create a poor appearance. The whole idea is to put your best foot forward to indicate you are the most enthusiastic and best-qualified candidate for the job. Contrary to what some applicants may think, outward appearances are very important. For the limited amount of time an interview board spends with a test applicant, all things become relevant, including the smallest of details.

The interview itself is normally conducted by a board of three or four people. Most interview panels consist of fire department officials or civil-service personnel. Occasionally, people outside the fire department or civil service are brought in to avoid potential bias on the board.

Ideally, those conducting the interview and the applicant being interviewed are complete strangers to one another. This way, if a candidate is not hired, he or she cannot discredit the selection process on the basis of bias or favoritism. Board members are also made aware that race, sex, color, creed, or political background has no bearing on these proceedings. Each interviewer has a rating sheet listing specific qualifications. The series of questions provides the interviewers enough insight to accurately gauge his or her potential capabilities. Usually the beginning of the interview will focus attention on your job application form. Such things as your educational background, past employment history, and references are examined. It would behoove you to review everything you listed on your application form and have supportive reasoning for any career changes. If you can somehow demonstrate that the direction you took was based on the underlying aspiration to become a firefighter, so much the better. However, try not to deceive the panel regarding past choices. Chances are that if you do, you may contradict yourself at one point or another and this will become immediately evident to the interviewers. The best policy here is to answer all questions honestly, even if some past decisions were not necessarily the best ones. If you feel that you have made a questionable career move or have had a falling out with one or more past employers, explain why. If you can also show that something was learned or gained from the experience, point that out as well. Interviewers will appreciate your honesty and sincerity. A history of switching jobs or changing careers all too frequently without just cause is usually reason enough not to be hired.

While you are being interviewed in these areas of concern, interviewers will be assessing your communication skills and how well you respond to the questioning. It is well understood that interviews are stressful to applicants. However, if an applicant appears excessively fidgety or worried or perspires profusely, and such nervousness encumbers the applicant's ability to answer questions, it can detract from what otherwise would have been a good interview. Advance preparation for the interview should help in this regard. Knowing in general the kinds of questions interviewers most likely will ask will enhance your confidence. Beside further expounding on information given on the application form, questions such as the following are equally important:

- Why do you want to become a firefighter?
- What merits you being hired over other similarly qualified applicants?
- Now that we know your strong points, what are your weaknesses? If you had to do everything over, what would you do differently? Do you have any regrets for something that happened in the past?
- What, if anything, do you feel are major accomplishments or achievements in your life?
- Is there any reason you did not actively participating in athletics in school?
- How do you feel you can help the community by working in the fire department?

These and a myriad of other questions are thought provoking. If you are prepared for such questioning, you will be better able to answer these questions in a satisfactory manner, rather than pausing at length to think of something. Simply answering "yes" or "no" is not sufficient. Supportive reasoning, even if it is brief, is what interviewers want to hear.

Concern during an interview will also focus on what your interests or hobbies are, as well as attitudes toward particular job requirements—for instance,

- Why do you like to hunt, swim, bike, camp, etc?
- Couldn't you have used your leisure time to more avail?
- Do you do any extracurricular reading and if so, what?
- Do you keep current with local events by reading the paper?
- How do you feel about working irregular hours in the fire department?
- Do you respond to criticism in a positive manner?
- Have you ever displayed temper with co-workers at past jobs?
- What do you think of drugs and alcohol, both in the work place and at home?
- Are you afraid of anything such as dying, heights, or speaking in front of large groups of people?

Having prepared answers to these questions and others of similar nature will definitely give you an edge over those who aren't prepared. Try to think of as many questions about your life as possible and prepare some reasoning to support your answers. You may be caught off guard by a few questions, but over all your preparation will pay off.

One other form the interview may take may concern your reaction to hypothetical circumstances or emergencies. It is not expected that you have advance knowledge of any specialized firefighter training. However, this kind of question can give interviewers insight into how well you can quickly reason and solve a problem. You may be given certain conditions to work within and then it is left entirely up to you to bring the situation under control.

These kinds of questions are obviously more difficult to prepare for, but two things are important to keep in mind. First, safety, to both fire personnel as well as the victim or members of the public involved, is a primary consideration. Second, nearly all effort made on behalf of fire personnel should be orchestrated as a team. Consider these two things during any questioning. Interviewers will describe some situation and may very well throw in some constraints that may make the situation worse. Whatever is given, think the question through as best you can and decide how you would handle the circumstances. Immediately answers to questions of this nature without much forethought is bound to be incomplete and show poor judgment. Interviewers will observe how tell you can assimilate information and identify specific problem areas. Your initiative and leadership beyond what is minimally necessary are other factors assessed as well.

If your interview is more in line with this kind of questioning, answer to the best of your ability and see the exercise through to the end. Whatever you do, don't become exasperated with the situation given and

THE INTERVIEW—THE FINAL PHASE OF SELECTION

give up. Remember, the interviewers know that they are placing you in a very stressful position. Reacting in an appropriate and confident manner bodes well in your employment consideration.

When the interview is winding to a close, one of the panel members will ask you if you have any questions or concerns regarding the fire department. If you feel that you have other positive qualities that were not discussed during the interview, now is the time to mention briefly what was overlooked. If you have some specific concerns regarding the fire department, this is the appropriate time to ask. Since there are other candidates to be interviewed, do not protract your own interview beyond a few minutes after they ask you for any further comments. Rambling on about something longer than necessary is viewed with disdain. Be brief with your questions if you have any then thank each interviewer for his or her time and consideration.

Don't loiter after the interview to see how well you did. It won't be for another one to two weeks before all things are considered and decisions are made with regard to hiring.

If you later learn you did not fare as well as expected in the interview, don't become upset and write the experience off as though the examiners made the mistake. Rather, find out where your weaknesses were and learn from that experience. That way on a follow-up interview to another exam, you will not make the same mistake. It can also be said that a candidate who goes through the testing and selection process more than once is very determined. That attribute is looked upon favorably by any fire department because it shows the applicant is truly dedicated to becoming a firefighter. More often than not, these are candidates that a fire department seek to hire. There may be a few disappointments along the way to being hired; however, hard work and persistence are two key virtues that are prerequisites to a fulfilling career in the fire department.

Other titles by Norman Hall

Money-Back guarantee! No other exam books make this offer because no other exam books are as comprehensive and up-to-date!

Corrections Officer Exam Preparation Book

Test expert Norman Hall shows readers guaranteed methods for scoring 80% to 100% on the corrections officer test. Hall analyzes every aspect of the most current version of the test and shows readers what they need to qualify, from memory tests to basic mathematics. Norman Hall covers everything you'll need to know to be hired, including:

- Written exams
- Physical abilities test
- Oral boards
- Psychological examinations
- And more

Careers, trade paperback, 8½" x 11", $10.95, 1-55850-793-0

State Trooper & Highway Patrol Exam Preparation Book

Guaranteed methods for scoring 80% to 100% on the state trooper and highway patrol officer qualification tests. Hall analyzes every aspect of the most current versions of the tests—from reading comprehension to simple math to physical fitness—and shows readers what they need to qualify including:

- Memory
- Reading Comprehension
- Reasoning and Judgment
- Map Reading
- Report Writing
- Grammar, Vocabulary, and Spelling

Careers, trade paperback, 8½" X 11", 296 pages, $12.95, 1-58062-077-9

Available wherever books are sold.

HOW TO ORDER: If you cannot find these titles at your favorite retail outlet, you may order them directly from the publisher. BY PHONE: Call 1-800-872-5627. We accept Visa, Mastercard, and American Express. $4.95 will be added to your total order for shipping and handling. BY MAIL: Write out the full titles of the books you'd like to order and send payment, including $4.95 for shipping and handling, to: Adams Media Corporation, 260 Center Street, Holbrook, MA 02343. 30-day money-back guarantee.

Other titles by Norman Hall

Money-Back guarantee! No other exam books make this offer because no other exam books are as comprehensive and up-to-date!

Postal Exam Preparation Book: Completely revised second edition

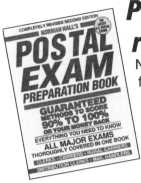

Norman Hall knows the techniques, winning strategies, and preparation necessary for the Postal Exam…and it's all inside and guaranteed to work for you! Everything you need to know to come out on top, complete coverage of:

- Address cross comparison
- Memory (names & numbers)
- Number series
- Following Directions
- Test-taking strategies and study suggestions

Careers, trade paperback, 8 ½" x 11", 296 pages, $10.95, 1-55850-363-3

Police Exam Preparation Book

Taking the exam to become a law enforcement officer can be challenging. Fortunately, if you use *The Complete Police Exam Preparation Book*, you won't have to worry about whether you'll come up with a positive performance on exam day.

Everything you need to know about a Career in Law Enforcement is thoroughly covered in one book. Complete coverage of ALL test subject areas. Everything you need to know to come out on top.

Norman Hall's *Complete Police Exam Preparation Book* includes:

- 7 practice tests covering key subject areas
- 2 full-length Police Officer Exams
- Answer keys and self-scoring tables
- Pointers on avoiding common trouble spots
- Meeting the physical requirements
- Plus: the latest test-taking strategies

Careers, trade paperback, 8 ½" x 11", 256 pages, $12.95, 1-55850-296-3

Available wherever books are sold.

HOW TO ORDER: If you cannot find these titles at your favorite retail outlet, you may order them directly from the publisher. BY PHONE: Call 1-800-872-5627. We accept Visa, Mastercard, and American Express. $4.95 will be added to your total order for shipping and handling. BY MAIL: Write out the full titles of the books you'd like to order and send payment, including $4.95 for shipping and handling, to: Adams Media Corporation, 260 Center Street, Holbrook, MA 02343. 30-day money-back guarantee.

Refund Policy

In the unlikely event that you use this book but score less than 80 percent on the firefighter's examination, your money will be refunded. This guarantee specifically applies to the written exam, not the physical fitness or medical exams: If a test applicant scores above 80 percent on the written test, but fails the physical fitness requirement or the medical exam, he or she will not be eligible for a refund.

The following conditions must be met before any refund will be made. All exercises in this guide must be completed to demonstrate that the applicant did make a real attempt to practice and prepare to score 80 percent or better. Any refund must be claimed within ninety days of the date of purchase shown on your sales receipt. Anything submitted beyond this ninety-day period will be subject to the publisher's discretion. Refunds are only available for copies of the book purchased through retail bookstores. The refund amount is limited to the purchase price and may not exceed the cover price of the book.

If you mail this study guide back for a refund, please include your sales receipt, validated test results,* and a self-addressed, stamped envelope. Requests for refunds should be addressed to Adams Media Corporation, Firefighter Exam Division, 260 Center Street, Holbrook, MA 02343. Please allow approximately four weeks for processing.

---

* On occasion, exam results are not mailed to the test applicant. Test scores may be posted at either the Fire Department or the place of examination. If this is the case for you, procure a copy of your test score from the personnel office and be sure your name and address are indicated (Social Security numbers are insufficient to claim a refund).